Hardware and Software Support for Virtualization

Synthesis Lectures on Computer Architecture

Editor
Margaret Martonosi, *Princeton University*

Synthesis Lectures on Computer Architecture publishes 50- to 100-page publications on topics pertaining to the science and art of designing, analyzing, selecting and interconnecting hardware components to create computers that meet functional, performance and cost goals. The scope will largely follow the purview of premier computer architecture conferences, such as ISCA, HPCA, MICRO, and ASPLOS.

Hardware and Software Support for Virtualization
Edouard Bugnion, Jason Nieh, and Dan Tsafrir
2017

Datacenter Design and Management: A Computer Architect's Perspective
Benjamin C. Lee
2016

A Primer on Compression in the Memory Hierarchy
Somayeh Sardashti, Angelos Arelakis, Per Stenström, and David A. Wood
2015

Research Infrastructures for Hardware Accelerators
Yakun Sophia Shao and David Brooks
2015

Analyzing Analytics
Rajesh Bordawekar, Bob Blainey, and Ruchir Puri
2015

Customizable Computing
Yu-Ting Chen, Jason Cong, Michael Gill, Glenn Reinman, and Bingjun Xiao
2015

Die-stacking Architecture
Yuan Xie and Jishen Zhao
2015

Single-Instruction Multiple-Data Execution
Christopher J. Hughes
2015

Power-Efficient Computer Architectures: Recent Advances
Magnus Själander, Margaret Martonosi, and Stefanos Kaxiras
2014

FPGA-Accelerated Simulation of Computer Systems
Hari Angepat, Derek Chiou, Eric S. Chung, and James C. Hoe
2014

A Primer on Hardware Prefetching
Babak Falsafi and Thomas F. Wenisch
2014

On-Chip Photonic Interconnects: A Computer Architect's Perspective
Christopher J. Nitta, Matthew K. Farrens, and Venkatesh Akella
2013

Optimization and Mathematical Modeling in Computer Architecture
Tony Nowatzki, Michael Ferris, Karthikeyan Sankaralingam, Cristian Estan, Nilay Vaish, and David
Wood
2013

Security Basics for Computer Architects
Ruby B. Lee
2013

The Datacenter as a Computer: An Introduction to the Design of Warehouse-Scale
Machines, Second edition
Luiz André Barroso, Jimmy Clidaras, and Urs Hölzle
2013

Shared-Memory Synchronization
Michael L. Scott
2013

Resilient Architecture Design for Voltage Variation
Vijay Janapa Reddi and Meeta Sharma Gupta
2013

Multithreading Architecture
Mario Nemirovsky and Dean M. Tullsen
2013

Performance Analysis and Tuning for General Purpose Graphics Processing Units (GPGPU)
Hyesoon Kim, Richard Vuduc, Sara Baghsorkhi, Jee Choi, and Wen-mei Hwu
2012

Automatic Parallelization: An Overview of Fundamental Compiler Techniques
Samuel P. Midkiff
2012

Phase Change Memory: From Devices to Systems
Moinuddin K. Qureshi, Sudhanva Gurumurthi, and Bipin Rajendran
2011

Multi-Core Cache Hierarchies
Rajeev Balasubramonian, Norman P. Jouppi, and Naveen Muralimanohar
2011

A Primer on Memory Consistency and Cache Coherence
Daniel J. Sorin, Mark D. Hill, and David A. Wood
2011

Dynamic Binary Modification: Tools, Techniques, and Applications
Kim Hazelwood
2011

Quantum Computing for Computer Architects, Second Edition
Tzvetan S. Metodi, Arvin I. Faruque, and Frederic T. Chong
2011

High Performance Datacenter Networks: Architectures, Algorithms, and Opportunities
Dennis Abts and John Kim
2011

Processor Microarchitecture: An Implementation Perspective
Antonio González, Fernando Latorre, and Grigorios Magklis
2010

Transactional Memory, 2nd edition
Tim Harris, James Larus, and Ravi Rajwar
2010

Computer Architecture Performance Evaluation Methods
Lieven Eeckhout
2010

Introduction to Reconfigurable Supercomputing
Marco Lanzagorta, Stephen Bique, and Robert Rosenberg
2009

Hardware and Software Support for Virtualization

Edouard Bugnion, Jason Nieh, and Dan Tsafrir

ISBN: 978-3-031-00625-8   paperback
ISBN: 978-3-031-01753-7   ebook

DOI 10.1007/978-3-031-01753-7

A Publication in the Springer series
SYNTHESIS LECTURES ON ADVANCES IN AUTOMOTIVE TECHNOLOGY

Lecture #38
Series Editor: Margaret Martonosi, *Princeton University*
Series ISSN
Print 1935-3235   Electronic 1935-3243

Hardware and Software Support for Virtualization

Edouard Bugnion
École Polytechnique Fédérale de Lausanne (EPFL), Switzerland

Jason Nieh
Columbia University

Dan Tsafrir
Technion – Israel Institute of Technology

SYNTHESIS LECTURES ON COMPUTER ARCHITECTURE #38

ABSTRACT

This book focuses on the core question of the necessary *architectural support provided by hardware* to efficiently run virtual machines, and of the corresponding design of the *hypervisors* that run them. Virtualization is still possible when the instruction set architecture lacks such support, but the hypervisor remains more complex and must rely on additional techniques.

Despite the focus on architectural support in current architectures, some historical perspective is necessary to appropriately frame the problem. The first half of the book provides the historical perspective of the theoretical framework developed four decades ago by Popek and Goldberg. It also describes earlier systems that enabled virtualization despite the lack of architectural support in hardware.

As is often the case, theory defines a necessary—but not sufficient—set of features, and modern architectures are the result of the combination of the theoretical framework with insights derived from practical systems. The second half of the book describes state-of-the-art support for virtualization in both x86-64 and ARM processors. This book includes an in-depth description of the CPU, memory, and I/O virtualization of these two processor architectures, as well as case studies on the Linux/KVM, VMware, and Xen hypervisors. It concludes with a performance comparison of virtualization on current-generation x86- and ARM-based systems across multiple hypervisors.

KEYWORDS

computer architecture, virtualization, virtual machine, hypervisor, dynamic binary translation

Contents

Preface

> "Virtual machines have finally arrived. Dismissed for a number of years as merely academic curiosities, they are now seen as cost-effective techniques for organizing computer systems resources to provide extraordinary system flexibility and support for certain unique applications".
>
> Robert. P. Goldberg, *IEEE Computer*, 1974 [78]

The academic discipline of computer systems research, including computer architecture, is in many aspects more tidal than linear: specific ingrained, well-understood techniques lose their relevance as tradeoffs evolve. Hence, the understanding of these techniques then ebbs from the collective knowledge of the community. Should the architectural tide later flow in the reverse direction, we have the opportunity to reinvent—or at least appreciate once more—old concepts all over again.

The history of virtualization is an excellent example of this cycle of innovation. The approach was popular in the early era of computing, as demonstrated from the opening quote. At high tide in the 1970s, hundreds of papers were written on virtualization with conferences and workshops dedicated to the topic. The era established the basic principles of virtualization and entire compute stacks—hardware, virtual machine monitors, and operating systems—were designed to efficiently support virtual machines. However, the tide receded quickly in the early 1980s as operating systems matured; virtual machines were soon strategically discarded in favor of a more operating system-centric approach to building systems.

Throughout the 1980s and 1990s, with the appearance of the personal computer and client/server era, virtual machines were largely relegated to a mainframe-specific curiosity. For example, the processors developed in that era (MIPS, Sparc, x86), were not explicitly designed to provide architectural support for virtualization, since there was no obvious business requirement to maintain support for virtual machines. In addition, and in good part because of the ebb of knowledge of the formal requirements for virtualization, many of these architectures made arbitrary design decisions that violated the basic principles established a decade earlier.

For most computer systems researchers of the open systems era, raised on UNIX, RISC, and x86, virtual machines were perceived to be just another bad idea from the 1970s. In 1997, the Disco [44] paper revisited virtual machines with a fresh outlook, specifically as the foundation to run commodity operating systems on scalable multiprocessors. In 1999, VMware released VMware Workstation 1.0 [45], the first commercial virtualization solution for x86 processors.

At the time, researchers and commercial entities started building virtual machines solutions for desktops and servers. A few years later, the approach was introduced to mobile platforms. Disco, VMware Workstation, VMware ESX Server [177], VirtualPC [130], Xen [27], Denali [182], and Cells [16], were all originally designed for architectures that did *not* provide support for virtualization. These different software systems each took a different approach to work around the limitations of the hardware of the time. Although processor architectures have evolved to provide hardware support for virtualization, many of the key innovations of that era such as hosted architectures [162], paravirtualization [27, 182], live migration [51, 135], and memory ballooning [177], remain relevant today, and have a profound impact on computer architecture trends.

Clearly, the virtualization tide has turned, to the point that it is once more a central driver of innovation throughout the industry, including system software, systems management, processor design, and I/O architectures. As a matter of fact, the exact quote from Goldberg's 1974 paper would have been equally timely 30 years later: Intel introduced its first-generation hardware support for virtual machines in 2004. Every maintained virtualization solution, including VMware Workstation, ESX Server, and Xen, quickly evolved to leverage the benefits of hardware support for virtualization. New systems were introduced that assumed the existence of such hardware support as a core design principle, notably KVM [113]. With the combined innovation in hardware and software and the full support of the entire industry, virtual machines quickly became central to IT organizations, where they were used among other things to improve IT efficiency, simplify provisioning, and increase availability of applications. Virtual machines were also proposed to uniquely solve hard open research questions, in domains such as live migration [51, 135] and security [73]. Within a few years, they would play a central role in enterprise datacenters. For example, according to the market research firm IDC, since 2009 there are more virtual machines deployed than physical hosts [95].

Today, virtual machines are ubiquitous in enterprise environments, where they are used to virtualize servers as well as desktops. They form the foundation of all Infrastructure-as-a-Service (IAAS) clouds, including Amazon EC2, Google CGE, Microsoft Azure, and OpenStack. Once again, the academic community dedicates conference tracks, sessions, and workshops to the topic (e.g., the annual conference on Virtual Execution Environments (VEE)).

ORGANIZATION OF THIS BOOK

This book focuses on the core question of the necessary *architectural support provided by hardware* to efficiently run virtual machines. Despite the focus on architectural support in current architectures, some historical perspective is necessary to appropriately frame the problem. Specifically, this includes both a theoretical framework, and a description of the systems enabling virtualization despite the lack of architectural support in hardware. As is often the case, theory defines a necessary—but not sufficient—set of features, and modern architectures are the result of the combination of the theoretical framework with insights derived from practical systems.

The book is organized as follows.

- Chapter 1 introduces the fundamental definitions of the abstraction ("virtual machines"), the run-time ("virtual machine monitors"), and the principles used to implement them.

- Chapter 2 provides the necessary theoretical framework that defines whether an instruction set architecture (ISA) is virtualizable or not, as formalized by Popek and Goldberg [143].

- Chapter 3 then describes the first set of systems designed for platforms that failed the Popek/Goldberg test. These systems each use a particular combination of workarounds to run virtual machines on platforms not designed for them. Although a historical curiosity by now, some of the techniques developed during that era remain relevant today.

- Chapter 4 focuses on the architectural support for virtualization of modern x86-64 processors, and in particular Intel's VT-x extensions. It uses KVM as a detailed case study of a hypervisor specifically designed to assume the presence of virtualization features in processors.

- Chapter 5 continues the description of x86-64 on the related question of the architectural support for MMU virtualization provided by extended page tables (also known as nested page tables).

- Chapter 6 closes the description of x86-64 virtualization with the various forms of I/O virtualization available. The chapter covers key concepts such as I/O emulation provided by hypervisors, paravirtual I/O devices, pass-through I/O with SR-IOV, IOMMUs, and the support for interrupt virtualization.

- Chapter 7 describes the architectural support for virtualization of the ARM processor family, and covers the CPU, MMU, and I/O considerations. The chapter emphasizes some of the key differences in design decisions between x86 and ARM.

- Chapter 8 compares the performance and overheads of virtualization extensions on x86 and on ARM.

In preparing this book, the authors made some deliberate decisions. First, for brevity, we focused on the examples of architectural support for virtualization, primarily around two architectures: x86-64 and ARM. Interested readers are hereby encouraged to study additional instruction set architectures. Among them, IBM POWER architecture, with its support for both hypervisor-based virtualization and logical partitioning (LPAR), is an obvious choice [76]. The SPARC architecture also provides built-in support for logical partitioning, called logical domains [163]. We also omit any detailed technical description of mainframe and mainframe-era architectures. Readers interested in that topic should start with Goldberg's survey paper [78] and Creasy's overview of the IBM VM/370 system [54].

Second, we focused on mainstream (i.e., traditional) forms of virtual machines and the construction of hypervisors in both the presence or the absence of architectural support for virtualization in hardware. This focus is done at the expense of a description of some more advanced research concepts. For example, the text does not discuss recursive virtual machines [33, 158], the use of virtualization hardware for purposes other than running traditional virtual machines [24, 29, 31, 43, 88], or the emerging question of architectural support for containers such as Docker [129].

AUTHORS' PERSPECTIVES

This book does not attempt to cover all aspects of virtualization. Rather, it mostly focuses on the key question of the interaction between the underlying computer architecture and the systems software built on top of it. It also comes with a point of view, based on the authors' direct experiences and perspectives on the topic.

Edouard Bugnion was fortunate to be part of the Disco team as a graduate student. Because of the stigma associated with virtual machines of an earlier generation, we named our prototype in reference to the questionable musical contribution of that same decade [55], which was then coincidentally making a temporary comeback. Edouard later co-founded VMware, where he was one of the main architects and implementers of VMware Workstation, and then served as its Chief Technology Officer. In 2005, he co-founded Nuova Systems, a hardware company premised on providing architectural support for virtualization in the network and the I/O subsystem, which became the core of Cisco's Data Center strategy. More recently, having returned to academia as a professor at École polytechnique fédérale de Lausanne (EPFL), Edouard is now involved in the IX project [30, 31, 147] which leverages virtualization hardware and the Dune framework [29] to build specialized operating systems.

Jason Nieh is a Professor of Computer Science at Columbia University, where he has led a wide range of virtualization research projects that have helped shape commercial and educational practice. Zap [138], an early lightweight virtual machine architecture that supported migration, led to the development of Linux namespaces and Linux containers, as well as his later work on Cells [16, 56], one of the first mobile virtualization solutions. Virtual Layered File Systems [144, 145] introduced the core ideas of layers and repositories behind Docker and CoreOS. KVM/ARM [60] is widely deployed and used as the mainline Linux ARM hypervisor, and has led to improvements in ARM architectural support for virtualization [58]. MobiDesk [26], THINC [25], and other detailed measurement studies helped make the case for virtual desktop infrastructure, which has become widely used in industry. A dedicated teacher, Jason was the first to introduce virtualization as a pedagogical tool for teaching hands-on computer science courses, such as operating systems [136, 137], which has become common practice in universities around the world.

Dan Tsafrir is an Associate Professor at the Technion—Israel Institute of Technology, where he regularly appreciates how fortunate he is to be working with brilliant students on cool projects for a living. Some of these projects drive state-of-the-art virtualization forward. For example, vIOMMU showed for the first time how to fully virtualize I/O devices on separate (side)cores without the knowledge or involvement of virtual machines, thus eliminating seemingly inherent trap-and-emulate virtualization overheads [12]. vRIO showed that such sidecores can in fact be consolidated on separate remote servers, enabling a new kind of datacenter-scale I/O virtualization model that is cheaper and more performant than existing alternatives [116]. ELI introduced software-based exitless interrupts—a concept recently adopted by hardware—which, after years of efforts, finally provided bare-metal performance for high-throughput virtualization workloads [13, 80]. VSwapper showed that uncooperative swapping of memory of virtual machines can be made efficient, despite the common belief that this is impossible [14]. Virtual CPU validation showed how to uncover a massive amount of (confirmed and now fixed) hypervisor bugs by applying Intel's physical CPU testing infrastructure to the KVM hypervisor [15]. EIOVAR and its successor projects allowed for substantially faster and safer IOMMU protection and found their way into the Linux kernel [126, 127, 142]. NPFs provide page-fault support for network controllers and are now implemented in production Mellanox NICs [120].

TARGET AUDIENCE

This book is written for researchers and graduate students who have already taken a basic course in both computer architecture and operating systems, and who are interested in becoming fluent with virtualization concepts. Given the recurrence of virtualization in the literature, it should be particularly useful to new graduate students before they start reading the many papers treating a particular sub-aspect of virtualization. We include numerous references of widely read papers on the topic, together with a high-level, modern commentary on their impact and relevance today.

Edouard Bugnion, Jason Nieh, and Dan Tsafrir
January 2017

Acknowledgments

This book would not have happened without the support of many colleagues. The process would have not even started without the original suggestion from Rich Uhlig to Margaret Martonosi, the series editor. The process, in all likelihood, would have never ended without the constant, gentle probing of Mike Morgan; we thank him for his patience. Ole Agesen, Christoffer Dall, Arthur Kiyanovski, Shih-Wei Li, Jintack Lim, George Prekas, Jeff Sheldon, and Igor Smolyar provided performance figures specifically for this book; students will find the additional quantitative data enlightening. Margaret Church made multiple copy-editing passes to the manuscript; we thank her for the diligent and detailed feedback at each round. Nadav Amit, Christoffer Dall, Nathan Dauthenhahn, Canturk Isci, Arthur Kiyanovski, Christos Kozyrakis, Igor Smolyar, Ravi Soundararajan, Michael Swift, and Idan Yaniv all provided great technical feedback on the manuscript.

The authors would like to thank EPFL, Columbia University, and the Technion—Israel Institute of Technology, for their institutional support. Bugnion's research group is supported in part by grants from Nano-Tera, the Microsoft EPFL Joint Research Center, a Google Graduate Research Fellowship and a VMware research grant. Nieh's research group is supported in part by ARM Ltd., Huawei Technologies, a Google Research Award, and NSF grants CNS-1162447, CNS-1422909, and CCF-1162021. Tsafrir's research group is supported in part by research awards from Google Inc., Intel Corporation, Mellanox Technologies, and VMware Inc., as well as by funding from the Israel Science Foundation (ISF) grant No. 605/12, the Israeli Ministry of Economics via the HIPER consortium, the joint BSF-NSF United States-Israel Binational Science Foundation and National Science Foundation grant No. 2014621, and the European Union's Horizon 2020 research and innovation programme grant agreement No. 688386 (OPERA).

Edouard Bugnion, Jason Nieh, and Dan Tsafrir
Lausanne, New York, and Haifa
January 2017

CHAPTER 1

Definitions

This chapter introduces the basic concepts of virtualization, virtual machines, and virtual machine monitors. This is necessary for clarity as various articles, textbooks, and commercial product descriptions sometimes use conflicting definitions. We use the following definitions in this book.

- **Virtualization** is the application of the layering principle through enforced modularity, whereby the exposed virtual resource is identical to the underlying physical resource being virtualized.

- A **virtual machine** is an abstraction of a complete compute environment through the combined virtualization of the processor, memory, and I/O components of a computer.

- The **hypervisor** is a specialized piece of system software that manages and runs virtual machines.

- The **virtual machine monitor (VMM)** refers to the portion of the hypervisor that focuses on the CPU and memory virtualization.[1]

The rest of this chapter is organized as follows. We formalize the definitions of virtualization, virtual machines, and hypervisors in §1.1, §1.2, and §1.3, respectively. §1.4 classifies existing hypervisors into type-1 (bare-metal) and type-2 (hosted) architectures. We then provide a sketch illustration of a hypervisor in §1.5 while deferring the formal definition until Chapter 2. In addition to the three basic concepts, we also define useful, adjacent concepts such as the different terms for memory (§1.6) and various approaches to virtualization and paravirtualization (§1.7). We conclude in §1.8 with a short description of some of the key reasons why virtual machines play a fundamental role in information technology today.

1.1 VIRTUALIZATION

> **Virtualization** is the application of the layering principle through enforced modularity, whereby the exposed virtual resource is identical to the underlying physical resource being virtualized.

---

[1]N.B.: the terms hypervisor and virtual machine monitor have been used interchangeably in the literature. Here, we prefer the term *hypervisor* when describing an entire system and the term *VMM* when describing the subsystem that virtualizes the CPU and memory, or in its historical formal context in Chapter 2.

This definition is grounded in two fundamental principles of computer systems. First, **layering** is the presentation of a single abstraction, realized by adding a level of indirection, when (i) the indirection relies on a single lower layer and (ii) uses a well-defined namespace to expose the abstraction. Second, **enforced modularity** additionally guarantees that the clients of the layer cannot bypass the abstraction layer, for example to access the physical resource directly or have visibility into the usage of the underlying physical namespace. Virtualization is therefore nothing more than an instance of layering for which the exposed abstraction is equivalent to the underlying physical resource.

This combination of indirection, enforced modularity, and compatibility is a particularly powerful way to both reduce the complexity of computer systems and simplify operations. Let's take the classic example of RAID [48], in which a redundant array of inexpensive disk is aggregated to form a single, virtual disk. Because the interface is compatible (it is a block device for both the virtual and physical disks), a filesystem can be deployed identically, whether the RAID layer is present or not. As the RAID layer manages its own resources internally, hiding the physical addresses from the abstraction, physical disks can be swapped into the virtual disk transparently from the filesystem using it; this simplifies operations, in particular when disks fail and must be replaced. Even though RAID hides many details of the organization of the storage subsystem from the filesystem, the operational benefits clearly outweigh any potential drawbacks resulting from the added level of indirection.

As broadly defined, virtualization is therefore not synonymous to virtual machines. It is also not limited to any particular field of computer science or location in the compute stack. In fact, virtualization is prevalent across domains. We provide a few examples found in hardware, software, and firmware.

Virtualization in Computer Architecture: Virtualization is obviously a fundamental part of computer architecture. Virtual memory, as exposed through memory management units (MMU), serves as the canonical example: the MMU adds a level of indirection which hides the physical addresses from applications, in general through a combination of segmentation and paging mechanisms. This enforces modularity as MMU control operations are restricted to kernel mode. As both physical memory and virtual memory expose the same abstraction of byte-addressable memory, the same instruction set architecture can operate identically with virtual memory when the MMU is enabled, and with physical memory when it is disabled.

Virtualization within Operating Systems: Operating systems have largely adopted the same concept. In fact, at its core, an operating system does little more than safely expose the resources of a computer—CPU, memory, and I/O—to multiple, concurrent applications. For example, an operating system controls the MMU to expose the abstraction of isolated address spaces to processes; it schedules threads on the physical cores transparently, thereby multiplexing in software the limited physical CPU resource; it mounts multiple distinct filesystems into a single virtualized namespace.

Virtualization in I/O subsystems: Virtualization is ubiquitous in disks and disk controllers, where the resource being virtualized is a block-addressed array of sectors. The approach is used by RAID controllers and storage arrays, which present the abstraction of multiple (virtual) disks to the operating systems, which addresses them as (real) disks. Similarly, the Flash Translation Layer found in current SSD provides wear-leveling within the I/O subsystem and exposes the SSD to the operating systems as though it were a mechanical disk.

Whether done in hardware, in software, or embedded in subsystems, virtualization is always achieved by using and combining three simple techniques, illustrated in Figure 1.1. First, **multiplexing** exposes a resource among multiple virtual entities. There are two types of multiplexing, in space and in time. With space multiplexing, the physical resource is partitioned (in space) into virtual entities. For example, the operating system multiplexes different pages of physical memory across different address spaces. To achieve this goal, the operating system manages the virtual-to-physical mappings and relies on the architectural support provided by the MMU.

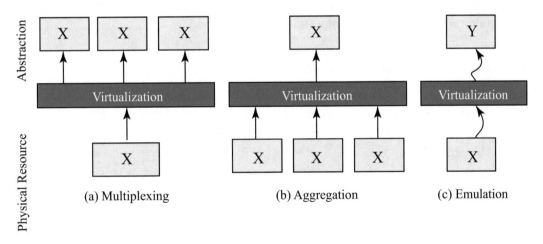

Figure 1.1: Three basic implementations techniques of virtualization. X represents both the physical resource and the virtualized abstraction.

With time multiplexing, the same physical resource is scheduled temporally between virtual entities. For example, the OS scheduler multiplexes the CPU core and hardware threads among the set of runnable processes. The context switching operation saves the processor's register file in the memory associated with the outgoing process, and then restores the state of the register file from the memory location associated with the incoming process.

Second, **aggregation** does the opposite, it takes multiple physical resources and makes them appear as a single abstraction. For example, a RAID controller aggregates multiple disks into a single volume. Once configured, the controller ensures that all read and write operations to the volume are appropriately reflected onto the various disks of the RAID group. The operating system then formats the filesystem onto the volume without having to worry about the details of the layout and the encoding. In a different domain, a processor's memory controller aggregates the capacity of multiple DIMMs into a single physical address space, which is then managed by the operating system.

Third, **emulation** relies on a level of indirection in software to expose a virtual resource or device that corresponds to a physical device, even if it is not present in the current computer system. Cross-architectural emulators run one processor architecture on another, e.g., Apple Rosetta emulates a PowerPC processor on an x86 computer for backward compatibility. In this example, X=PowerPC and Y=x86 in Figure 1.1c. The virtual abstraction corresponds to a particular processor with a well-defined ISA, even though the physical processor is different. Memory and disks can emulate each other: a RAM disk emulates the function of a disk using DRAM as backing store. The paging process of virtual memory does the opposite: the operating system uses disk sectors to emulate virtual memory.

Multiplexing, aggregation, and emulation can naturally be combined together to form a complete execution stack. In particular, as we will see shortly in §1.5, nearly all hypervisors incorporate a combination of multiplexing and emulation.

1.2 VIRTUAL MACHINES

The term "virtual machine" has been used to describe different abstractions depending on epoch and context. Fortunately, all uses are consistent with the following, broad definition.

> A **virtual machine** is a complete compute environment with its own isolated processing capabilities, memory, and communication channels.

This definition applies to a range of distinct, incompatible abstractions, illustrated in Figure 1.2:

- **language-based virtual machines**, such as the Java Virtual Machine, Microsoft Common Language Runtime, Javascript engines embedded in browsers, and in general the run-time environment of any managed language. These runtime environments are focused on running single applications and are not within the scope of this book;

- **lightweight virtual machines**, which rely on a combination of hardware and software isolation mechanisms to ensure that applications running directly on the processor (e.g., as native x86 code) are securely isolated from other sandboxes and the underlying operating system. This includes server-centric systems such as Denali [182] as well as desktop-centric systems such as the Google Native Client [190] and Vx32 [71]. Solutions based on Linux containers such as Docker [129] or the equivalent FreeBSD Jail [110] fall into the same category. We will refer to some of these systems as applicable, in particular in the context of the use of particular processor features; and

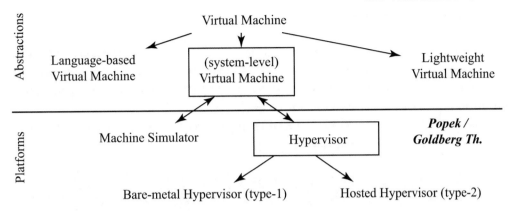

Figure 1.2: Basic classification of virtual machines and the platforms that run them.

- **system-level virtual machines**, in which the isolated compute environment resembles the hardware of a computer so that the virtual machine can run a standard, commodity operating system and its applications, in full isolation from the other virtual machines and the rest of the environment. Such virtual machines apply the virtualization principle to an *entire computer system*. Each virtual machine has its own copy of the underlying hardware, or at least, its own copy of *some* underlying hardware. Each virtual machine runs its own independent operating system instance, called the **guest operating system**. This is the essential focus of this book.

Figure 1.2 also categories the various platforms that run system-level virtual machines. We call these platforms either a hypervisor or a machine simulator, depending on the techniques used to run the virtual machine:

- a **hypervisor** relies on **direct execution** on the CPU for maximum efficiency, ideally to eliminate performance overheads altogether. In direct execution, the hypervisor sets up the hardware environment, but then lets the virtual machine instructions execute directly on the processor. As these instruction sequences must operate within the abstraction of the virtual machine, their execution causes traps, which must be emulated by the hypervisor. This **trap-and-emulate** paradigm is central to the design of hypervisors; and

- a **machine simulator** is typically implemented as a normal user-level application, with the goal of providing an accurate simulation of the virtualized architecture, and often runs at a small fraction of the native speed, ranging from a 5× slowdown to a 1000× slowdown, depending on the level of simulation detail. Machine simulators play a fundamental role in computer architecture by allowing the detailed architectural study of complex workloads [36, 124, 153, 181].

1.3   HYPERVISORS

A **hypervisor** is a special form of system software that runs virtual machines with the goal of minimizing execution overheads. When multiple virtual machines co-exist simultaneously on the same computer system, the hypervisor multiplexes (i.e., allocates and schedules) the physical resources appropriately among the virtual machines.

Popek and Goldberg formalized the relationship between a virtual machine and hypervisor (which they call VMM) in 1974 as follows [143].

> A virtual machine is taken to be an **efficient, isolated duplicate** of the real machine. We explain these notions through the idea of a virtual machine monitor (VMM). As a piece of software, a VMM has three essential characteristics. First, the VMM provides an environment for programs which is essentially identical with the original machine; second, programs running in this environment show at worst only minor decreases in speed; and last, the VMM is in complete control of system resources.

Popek and Goldberg's definition is consistent with the broader definition of virtualization: the hypervisor applies the layering principle to the computer with three specific criteria of equivalence, safety and performance.

Equivalence: Duplication ensures that the exposed resource (i.e., the virtual machine) is equivalent with the underlying computer. This is a strong requirement, which has been historically relaxed in some measure when the architecture demands it (see §1.7).

Safety: Isolation requires that the virtual machines are isolated from each other as well as from the hypervisor, which enforces the modularity of the system. Critically, the safety of the design is enforced by the hypervisor without it making any assumptions about the software running inside the virtual machine (including the guest operating system).

Performance: Finally, and critically, Popek and Goldberg's definition adds an additional requirement: the virtual system must *show at worst a minor decrease in speed*. This final requirement separates hypervisors from machine simulators. Although machine simulators also meet the duplication and the isolation criteria, they fail the efficiency criteria as even fast machine simulators using dynamic binary translation [32, 124, 153, 184] slow down the execution of the target system by at least 5×, in large part because of the high cost of emulating the TLB of the virtual machine in software.

1.4   TYPE-1 AND TYPE-2 HYPERVISORS

Finally, Figure 1.2 shows that hypervisor architectures can be classified into so-called **type-1** and **type-2**. Robert Goldberg introduced these terms in his thesis [77], and the terms have been used ever since. Informally, a type-1 hypervisor is in direct controls of all resources of the physical computer. In contrast, a type-2 hypervisor operates either "as part of" or "on top of" an existing host operating system. Regrettably, the literature has applied the definitions loosely, leading to some confusion. Goldberg's definition (using an updated terminology) is as follows.

> The implementation requirement specifies that instructions execute directly on the host. It does not indicate how the hypervisor gains control for that subset of instructions that must be interpreted. This may be done either by a program running on the bare host machine or by a program running under some operating system on the host machine. In the case of running under an operating system, the host operating system primitives may be used to simplify writing the virtual machine monitor. Thus, two additional VMM categories arise:
>
> - **type-1:** the VMM runs on a bare machine;
>
> - **type-2:** the VMM runs on an extended host, under the host operating system.
>
> [...] In both type-1 and type-2 VMM, the VMM creates the virtual machine. However, in a type-1 environment, the VMM on a bare machine must perform the system's scheduling and (real) resource allocation. Thus, the type-1 VMM may include such code not specifically needed for virtualization. In a type-2 system, the resource allocation and environment creation functions for virtual machine are more clearly split. The operating system does the normal system resource allocation and provides a standard extended machine.

We note that the emphasis is on resource allocation, and not whether the hypervisor runs in privileged or non-privileged mode. In particular, a hypervisor can be a type-2 even when it runs in kernel-mode, e.g., Linux/KVM and VMware Workstation operate this way. In fact, Goldberg assumed that the hypervisor would always be executing with supervisor privileges.

Both types are commonly found in current systems. First, VMware ESX Server [177], Xen [27, 146], and Microsoft Hyper-V are all type-1 hypervisors. Even though Xen and Hyper-V depend on a host environment called *dom0*, the hypervisor itself makes the resource allocation and scheduling decisions.

Conversely, VMware Workstation [45], VMware Fusion, KVM [113], Microsoft Virtu-alPC, Parallels, and Oracle VirtualBox [179] are all type-2 hypervisors. They cooperate with a

host operating system so that the host operating system schedules all system resources. The host operating system schedules the hypervisor as if it were a process, even though these systems all depend on a heavy kernel-mode component to execute the virtual machine. Some hypervisors such as VMware Workstation and Oracle VirtualBox are portable across different host operating systems, while Fusion and Parallels runs with the Mac OS X host operating system, Microsoft Virtual PC runs with the Windows host operating system and KVM runs as part of the Linux host operating system. Among these type-2 systems, KVM provides the best integration with the host operating system, as the kernel mode component of the hypervisors is integrated directly within the Linux host as a kernel module.

1.5   A SKETCH HYPERVISOR: MULTIPLEXING AND EMULATION

With the basic definitions established, we now move to a first sketch description of the key elements of a virtualized computer system, i.e., the specification of a virtual machine and the basic building blocks of the hypervisor.

Figure 1.3 illustrates the key architectural components of a virtualized computer system. The figure shows three virtual machines, each with their own virtual hardware, their own guest operating system, and their own applications. The hypervisor controls the actual physical resources and runs directly on the hardware. In this simplified architecture, the hardware (virtual or physical) consists of processing elements, which comprises one or more CPUs, their MMU, and cache-coherent memory. The processing elements are connected to an I/O bus, with two attached I/O devices: a disk and a network interface card in Fig 1.3. This is representative of a server deployment. A desktop platform would include additional devices such as a keyboard, video, mouse, serial ports, USB ports, etc. A mobile platform might further require a GPS, an accelerometer, and radios.

In its most basic form, a hypervisor uses two of the three key virtualization techniques of §1.1: it *multiplexes* (in space, and possibly in time) the physical PE across the virtual machines, and it *emulates* everything else, in particular the I/O bus and the I/O devices.

This combination of techniques is both necessary and sufficient in practice to achieve the efficiency criteria. It is necessary because without an effective mechanism to multiplex the CPU and the MMU, the hypervisor would have to emulate the execution of the virtual machine. In fact, the principal difference between a machine simulator and a hypervisor is that the former emulates the virtual machine's instruction set architecture, while the later multiplexes it. Multiplexing of the CPU is a scheduling task, very similar to the one performed by the operating system to schedule processes. The scheduling entity (here, the hypervisor) sets up the hardware environments (register file, etc.) and then lets the scheduled entity (the virtual machine) run directly on the hardware with reduced privileges.

This scheduling technique is known as **direct execution** since the hypervisor lets the virtual CPU directly execute instructions on the real processor. Of course, the hypervisor is also

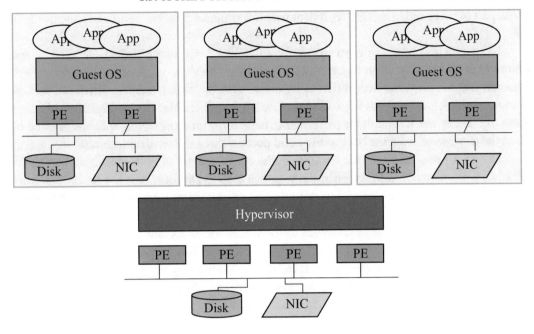

Figure 1.3: Basic architecture of a virtual machine monitor. Each processing element (PE) consists of CPU and physical memory.

responsible to ensure the safety property of the virtual machine. It therefore ensures that the virtual CPU always executes with reduced privileges, e.g., so that it cannot execute privileged instructions. As a consequence, the direct execution of the virtual machine leads to frequent traps whenever the guest operating system attempts to execute a privileged instruction, which must be emulated by the hypervisor. Hypervisors designed around direct execution therefore follow a **trap-and-emulate** programming paradigm, where the bulk of the execution overhead is due to the hypervisor emulating traps on behalf of the virtual machine.

Physical memory is also multiplexed among the virtual machines, so that each has the illusion of a contiguous, fixed-size, amount of physical memory. This is similar to the allocation among processes done by an operating system. The unique challenges in building a hypervisor lie in the virtualization of the MMU, and in the ability to expose user-level and kernel-level execution environment to the virtual machine.

The combination of multiplexing and emulation is also sufficient, as I/O operations of today's computer systems are implemented via reasonably high-level operations, e.g., a device driver can issue simple commands to send a list of network packets specified in a descriptor ring, or issue a 32 KB disk request. A hypervisor emulates the hardware/software interface of at least one representative device per category, i.e., one disk device, one network device, one screen device, etc. As part of this emulation, the hypervisor uses the available physical devices to issue the actual I/O.

I/O emulation has long been the preferred approach to the virtualization of I/O device because of its portability advantages: a virtual machine "sees" the same virtual hardware, even when running on platform with different hardware devices. Today, modern hardware includes advanced architectural support for I/O virtualization which enables the multiplexing of certain classes of I/O devices, with notable performance benefits in terms of throughput and latency, but still at the expense of reducing the mobility and portability of the virtual machines.

Table 1.1 provides a concrete example of the combined use of multiplexing and emulation in VMware Workstation 2.0, an early, desktop-oriented hypervisor, released in 2000. Clearly, the hardware is dated: USB is notably missing, and most readers have never seen actual floppy disks. The concepts however remain the same. For each component, Table 1.1 describes the **front-end device** abstraction, visible as hardware to the virtual machine, and the **back-end emulation** mechanisms used to implement it. When a resource is multiplexed, such as the x86 CPU or the memory, the front-end and back-end are identical, and defined by the hardware. The hypervisor is involved only to establish the mapping between the virtual and the physical resource, which the hardware can then directly use without further interception.

Table 1.1: Virtual Hardware of early VMware Workstation [45]

	Virtual Hardware (front-end)	Back-end
Multiplexed	1 virtual x86-32 CPU	Scheduled by the host operating system with one or more x86 CPUs
	Up to 512 MB of contiguous DRAM	Allocated and managed by the host OS (page-by-page)
Emulated	PCI Bus	Fully emulated compliant PCI bus with B/D/F addressing for all virtual mother-board and slot devices
	4 x 4IDE disks 7 x Buslogic SCSI Disks	Either virtual disks (stored as files) or direct access to a given raw device
	1 x IDE CD-ROM	ISO image or real CD-ROM
	2 x 1.44 MB floppy drives	Physical floppy or floppy image
	1 x VGA/SVGA graphics card	Appears as a Window or in full-screen mode
	2 x serial ports COM1 and COM2	Connect to Host serial port or a file
	1 x printer (LPT)	Can connect to host LPT port
	1 x keyboard (104-key) and mouse	Fully emulated
	AMD PCnet NIC (AM79C970A)	Via virtual switch of the host

When the resource is emulated, however, the **front-end device** corresponds to one canonically chosen representative of the device, independent of the back-end. The hypervisor implements both the front-end and the back-end, typically in software without any particular hardware support. The front-end is effectively a software model of the chosen, representative device. The **back-end emulation** chooses among the underlying resources to implement the functionality. These underlying resources may be physical devices or some higher-level abstractions of the host operating system. For example, the disk front-ends in VMware Workstation were either IDE or Buslogic SCSI devices, two popular choices at the time, with ubiquitous device drivers. The backend resource could be either a physical device, i.e., an actual raw disk, or a virtual disk stored as a large file within an existing filesystem.

Although no longer an exact duplicate of the underlying hardware, the virtual machine remains compatible. Assuming that a different set of device drivers can be loaded inside the guest operating system, the virtual machine will have the same functionality.

So far, this hypervisor sketch assumes that the various components of the processing elements can be virtualized. Yet, we've also alluded to the historical fact that some hardware architectures fail to provide hardware support for virtualization. This discussion will be the core of Chapter 2 and Chapter 3.

1.6   NAMES FOR MEMORY

The cliché claims (apparently incorrectly according to linguists) that Eskimos have many names for snow. Similarly, computer architecture and system designers have used at times overlapping, somewhat confusing definitions for the many facets of memory. The reason is simple. Like snow to Eskimos, virtual memory is fundamental to operating systems and arguably the most significant enhancements over the original von Neuman model of computing in this context. In reality, there are fundamentally only two types of memory: **virtual memory** converts into **physical memory** via the combination of segmentation and paging.

Virtual memory: Virtual memory refers to the byte-addressable namespace used by instruction sequences executed by the processor. With rare exceptions, all registers and the instruction pointer that refer to a memory location contain a **virtual address**. In a segmented architecture, the virtual address space is determined by a base address and a limit. The former is added to the virtual address and the latter is checked for protection purposes. In a paging architecture, the virtual address space is determined by the memory management unit on a page-by-page basis, with the mapping defined either by page tables or by a software TLB miss handler. Some architectures such as x86-32 combine segmentation with paging. The virtual address is first converted (via segmentation) into a **linear address**, and then (via paging) into a physical address.

Physical memory: Physical memory refers to the byte-addressable resource accessed via the memory hierarchy of the processor, and typically backed by DRAM. In a non-virtualized computer system, the **physical address space** is generally determined by the resources of the hardware, and defined by the memory controller of the processor. In a virtualized computer system, the hypervisor defines the amount of physical memory that is available to the virtual machine. Technically, the abstraction visible to the virtual machine should be called "virtual physical memory." However, to avoid confusion with virtual memory, it is most commonly called **guest-physical memory**. To further avoid ambiguity, the underlying resource is called **host-physical memory**. Note that some early virtualization papers used different terminologies, in particular using the terms *physical memory* and **machine memory** to refer to guest-physical and host-physical memory, respectively.

1.7   APPROACHES TO VIRTUALIZATION AND PARAVIRTUALIZATION

The lack of clear architectural support for virtualization in earlier processors has led to a range of approaches that solve the identical problem of running a virtual machine that is either similar or compatible with the underlying hardware. This text covers the three pragmatic approaches to virtualization.

Full (software) virtualization: This refers to hypervisors designed to maximize hardware compatibility, and in particular run unmodified operating systems, on architectures lacking the full support for it. It includes in particular early versions of VMware's hypervisors. This is also referenced as software virtualization in some papers. We will describe the approach in §3.2.

Hardware Virtualization (HVM):   This refers to hypervisors built for architectures that provide architectural support for virtualization, which includes all recent processors. Such hypervisors also support unmodified guest operating systems. Unlike software virtualization approach in which the hypervisor must (at least some of the time) translate guest instruction sequences before execution, HVM hypervisors rely exclusively on **direct execution** to execute virtual machine instructions. In the literature, HVM is at times referred to as HV. The requirements for an HVM are formally defined in Chapter 2. Architectures and hypervisors that follow the approach are the focus of Chapter 4 for x86 with VT-x and Chapter 7 for ARM with Virtualization Extensions.

Paravirtualization: This approach makes a different trade-off, and values simplicity and overall efficiency over the full compatibility with the underlying hardware. The term was introduced by Denali [182] and popularized by the early Xen hypervisor. In its original usage on platforms with no architectural support for virtualization, paravirtualization required a change of the guest operating system binary, which was incompatible with the underlying hardware. In its contemporary use on architectures with full virtualization support, paravirtualization is still used to augment the HVM through platform-specific extensions often implemented in device drivers, e.g., to manage cooperatively memory or to implement a high-performance front-end device. We will describe this in §3.3.

1.8   BENEFITS OF USING VIRTUAL MACHINES

So far, we have discussed the "how" of virtualization as any good engineering textbook should, but not the "why?", which is equally relevant. Virtual machines were first invented on mainframes when hardware was scarce and very expensive, and operating systems were primitive. Today, they are used very broadly because of fundamentally different considerations. We enumerate some of the most popular ones.

Operating system diversity: Virtualization enables a diversity of operating systems to run concurrently on a single machine. This benefit is the key reason behind many desktop-oriented type-2 hypervisors, e.g., Fusion and Parallels are essentially used to run Windows (and sometimes Linux) on Mac OS X.

Server consolidation: Enterprise IT best practices today still mandate that each server runs a single application per machine. With the rise in power and efficiency of hardware, that machine is today more often than not a virtual machine [95].

Rapid provisioning: Deploying a physical server is complex, time-intensive task. In contrast, a virtual machine can be created entirely in software through a portal or an API, and software stack can be deployed as virtual appliances [157].

Security: Virtualization introduces a new level of management in the datacenter stack, distinct and invisible from the guest operating systems, and yet capable of introspecting the behavior of such operating systems [49], performing intrusion analysis [68], or attesting of its provenance [73]. The hypervisor can also control all I/O operations from the virtual machine, making it easy to insert e.g., a VM-specific firewall or connect into a virtual network [114].

High-availability: A virtual machine is an encapsulated abstraction that can run on any server (running a compatible hypervisor). Specifically, a virtual machine can reboot following a hardware crash on a new server without operational impact, therefore providing a high-availability solution without requiring guest operating system-level configuration or awareness.

Distributed resource scheduling: The use of live migration technologies turns a cluster of hypervisors into a single resource pools, allowing the automatic and transparent rebalancing of virtual machines within the cluster [96].

Cloud computing: In a virtualized environment, different customers (tenants) can operate their own virtual machines in isolation from each other. When coupled with network virtualization (a technology outside of the scope of this text), this provides the foundation for the cloud computing technologies, including the ones of Amazon Web Services, Google Compute Engine, and Microsoft Azure.

1.9   FURTHER READING

We rely largely on Salzer and Kaashoek's book, *Principles of Computer Systems* [155], for definitions of layering and enforced modularity. The book provides excellent further reading to readers interested in additional background material, or who wish to look at other examples of virtualization. Tanenbaum and Bos' *Modern Operating Systems*, 4th ed. [165] also dedicates a chapter to virtualization.

CHAPTER 2

The Popek/Goldberg Theorem

In 1974, Gerald Popek and Robert Goldberg published in *Communications of the ACM* the seminal paper "Formal Requirements for Virtualizable Third-Generation Architectures" that defines the necessary and sufficient formal requirements to ensure that a VMM can be constructed [143]. Precisely, their theorem determines whether a given **instruction set architecture (ISA)** can be virtualized by a VMM using multiplexing. For any architecture that meets the hypothesis of the theorem, any operating system directly running on the hardware can also run inside a virtual machine, without modifications.

At the time, the motivation for the work was to address the evidence that new architectures accidentally prevented the construction of a VMM. The authors cited the DEC PDP-10 in their article, where seemingly arbitrary architectural decisions "broke" virtualization. Despite the simplicity of the result, the relevance of the theorem was lost on computer architects for decades, and generations of new processor architectures were designed without any technical considerations for the theorem.

Much later, as virtual machines once again became relevant, Intel and AMD explicitly made sure that their virtualization extensions met the Popek and Goldberg criteria, so that unmodified guest operating systems could run directly in virtual machines, without having to resort to software translation or paravirtualization [171].

Today, the theorem remains the obvious starting point to understand the fundamental relationship between a computer architecture and its ability to support virtual machines. Specifically, the theorem determines whether a VMM, relying exclusively on direct execution, can support guest arbitrary operating systems.

2.1 THE MODEL

The paper assumes a standard computer architecture, which the authors call a **conventional third-generation architecture**. The processor has two execution modes (user-level and supervisor), and support for virtual memory. Such an architecture is both necessary and sufficient to run a conventional operating system. In particular, the operating system can configure the hardware to run multiple, arbitrary, potentially malicious applications in isolation from each other. For the purpose of the proof, the paper defines a simple model that remains representative of the broader class of these architectures. Specifically, the model has the following features.

- The computer has one processor with two execution levels: supervisor mode and user mode.

- Virtual memory is implemented via segmentation (rather than paging) using a single segment with base B and limit L (called the relocation-bound register pair in the paper). The segment defines the range $[B, B + L[$ of the linear address space for valid virtual addresses $[0, L[$. There is no paging, so linear addresses map 1:1 directly to physical addresses. Virtual memory is used in both supervisor and user mode for all memory accesses.

- Physical memory is contiguous, starting at address 0, and the amount of physical memory is known at processor reset time (SZ).

- The processor's system state, called the **processor status word (PSW)** consists of the tuple (M, B, L, PC):

  - the execution level $M = \{s, u\}$;
  - the segment register (B,L); and
  - the current program counter (PC), a virtual address.

- The trap architecture has provisions to first save the content of the PSW to a well-known location in memory (MEM[0]), and then load into the PSW the values from another well-known location in memory (MEM[1]). The trap architecture mechanism is used to enter into the operating system following a system call or an exception in executing an instruction.

- The ISA includes at least one instruction or instruction sequence that can load into the hardware PSW the tuple (M, B, L, PC) from a location in virtual memory. This is required to resume execution in user mode after a system call or a trap.

- I/O and interrupts are ignored to simplify the discussion.

Let's first consider what an operating system for such an architecture would look like (in the absence of a VMM).

1. The kernel would run in supervisor mode ($M = s$), and applications would always run in user mode ($M = u$).

2. During initialization, the kernel first sets the trap entry point: MEM[0] ← (M:s,B:0,L:SZ,PC:trap_en).

3. The kernel would allocate a contiguous range of physical memory for each application.

4. To launch or resume an application stored in physical memory at [B,B+L[and currently executing instruction PC, the operating system would simply load PSW ← (M:u,L,C,PC).

5. At the trap entry point (PC =trap_en), the kernel would first decode the instruction stored at MEM[1].PC to determine the cause of the trap, and then take appropriate action.

Although idealized, notably because of the lack of registers, this architectural model is not fundamentally different from the ones that we are all familiar with today. For such an architecture, Popek and Goldberg posed the following formal research question.

> Given a computer that meets this basic architectural model, under which precise conditions can a VMM be constructed, so that the VMM:
>
> - can execute one or more virtual machines;
>
> - is in complete control of the machine at all times;
>
> - supports arbitrary, unmodified, and potentially malicious operating systems designed for that same architecture; and
>
> - be efficient to show at worst a small decrease in speed?

The answer to this question determines whether a VMM can be constructed for a particular architecture, so that the resulting "virtual machine can be an *efficient, isolated duplicate* of the real machine." When the conditions are met, the theorem must therefore ensure compliance with the following three criteria.

Equivalence: The virtual machine is essentially identical to the underlying processor, i.e., a *duplicate* of the computer architecture. Any program running within the virtual machine, i.e., any guest operating system and application combination, should exhibit identical behavior as if that program had run directly on the underlying hardware, save for the possible exception of differences caused by timing dependencies, or the availability of resources such as the amount of physical memory.

Safety: The VMM must be in complete control of the hardware at all times, without making any assumptions about the software running inside the virtual machine. A virtual machine is *isolated* from the underlying hardware and operates as if it were running on a distinct computer. In addition, different virtual machines must be isolated from each other. The theorem and its proof focus on the first safety property. The second property can be achieved by simply ensuring that there is no shared state within the VMM between two virtual machines.

Performance: The *efficiency* requirement implies that the execution speed of the program in a virtualized environment is at worst a minor decrease over the execution time when run directly on the underlying hardware.

2.2   THE THEOREM

The first theorem itself is simple.

> Theorem 1 [143]: For any conventional third-generation computer, a virtual machine monitor may be constructed if the set of sensitive instructions for that computer is a subset of the set of privileged instructions.

The key observation is that the answer to Popek and Goldberg's research question depends on the classification of the instructions of the ISA. An instruction is **control-sensitive** if it can update the system state, **behavior-sensitive** if its semantics depend on the actual values set in the system state, or an **innocuous instruction** otherwise. An instruction is also **privileged** if it can only be executed in supervisor mode and causes a trap when attempted from user mode. The theorem holds when all control-sensitive and behavior-sensitive instructions are also privileged, i.e.,

$$\{control\text{-}sensitive\} \cup \{behavior\text{-}sensitive\} \subseteq \{privileged\}.$$

Proof by construction: A VMM can be constructed if the architecture meets the constraints of the theorem. With some caveats, the converse normally holds, i.e., a VMM cannot be constructed if the architecture does not meet the constraints of the theorem.

Figure 2.1 illustrates the construction. First, Figure 2.2 delineates the relationships between the three key forms of memory: the host-physical memory, the guest-physical memory, and the virtual memory. Then, Figure 2.2 illustrates how the processor's PSW, and in particular the execution level M and the relocation-bound pair, is configured to run in three distinct situations: when running the VMM, the guest operating system, and the application.

If the conditions of the theorem are met, the VMM operates as follows.

1. The VMM is the only software actually running in supervisor mode. The VMM reserves a portion of physical memory for its own code and data structures, which is never mapped into any virtual address space of any virtual machine.

2. The VMM allocates the guest physical memory for each virtual machine contiguously in the host physical address space. Each virtual machine has a fixed, given amount of memory. Those parameters are illustrated as *addr0* and *memsize* in Figure 2.2.

3. The VMM keeps in memory a copy of the system state of each virtual machine: the vPSW. Like its hardware counterpart, the vPSW consists of the virtual machine's execution level, segment register (B,L) and program counter.

4. The virtual machine resumes execution by loading from memory PSW $\leftarrow (M', B', L', PC')$, with:

   - $M' \leftarrow u$: the virtual machine always executes in user mode;

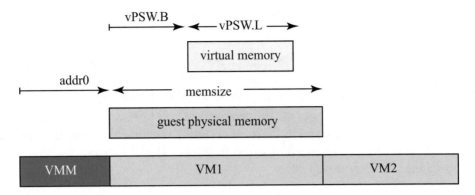

(a) Host physical, guest physical, and virtual memory.

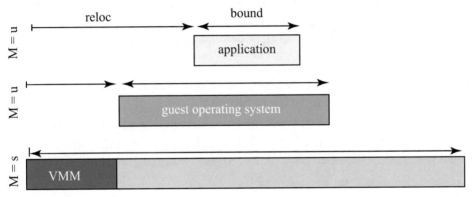

(b) Hardware configuration when executing applications, the guest operating system, and the VMM.

Figure 2.1: Construction of the the Popek/Goldberg VMM.

- $B' \leftarrow$ addr0+vPSW.B: the guest-physical offset is added to the virtual machine's segment base;

- $L' \leftarrow$ min(vPSW.L,vPSW.memsize-vPSW.B): in normal circumstances, the virtual machine's segment limit is used directly. However, a malicious guest operating system may attempt to allow its application to read beyond the end of guest-physical memory. The second clause of formula prevents this to ensure the safety property; and

- $PC' \leftarrow$ vPSW.PC: to resume execution of the virtual machine.

5. The VMM updates the PC (vPSW.PC $\leftarrow$ PSW.PC) on every trap. The other fields in the vPSW do no need updating even after an arbitrarily long direct execution period. Indeed, any instruction that can update the base, limit, or execution privilege level of the PSW is control-sensitive. The VMM assumes here that all control-sensitive instructions are also

privileged (the hypothesis of the theorem) and relies on the fact that the virtual machine always executes with $PSW.M \equiv u$).

6. The VMM then emulates the instruction that caused the trap. If the guest operating system was running ($vPSW.M \equiv s$), the VMM decodes the instruction and then emulates the semantic of the privileged instruction according to the ISA. For example, the VMM emulates the instruction that changes the segment register pair by updating $vPSW.B$ and $vPSW.L$ according to the new values set by the guest operating system. After a successful emulation, the VMM increments the program counter ($vPSW.PC{+}{+}$) and resumes execution of the virtual machine according to step #4.

7. The VMM may however conclude that the emulation of the instruction is actually a trap according to the ISA. For example, consider the case where the virtual machine was in user mode, (i.e., when $vPSW.M \equiv u$) and a malicious application attempted to issue a privileged instruction. Or perhaps if an application accesses memory beyond the limit set forth by the guest operating system in the segment descriptor. The VMM then emulates the trap according to the architecture, i.e.:

   - $MEM[addr0] \leftarrow vPSW$: store the virtual machine's vPSW into the first word in guest-physical memory, according to the architecture;

   - $vPSW \leftarrow MEM[addr0{+}1]$: copy the second word in guest-physical memory into the vPSW, according to the architecture. This updates the state of the virtual machine in line with the configuration set forth by the guest operating system; and

   - resume execution according to step #4.

8. If the hypothesis of the theorem is met, then all instructions that update the system state (other than the program counter) are control-sensitive, and therefore privileged. These will cause a trap. This would include any instruction that changes the base-bound registers defining the virtual address space (see above), or transitioning between supervisor and user mode. The VMM emulates such instructions according to the architecture specification.

9. If the hypothesis is met, any behavior-sensitive instruction would also be privileged, and therefore cause a trap from user mode. The VMM emulates these instructions as well. Consider for example an instruction that reads the execution level ($PSW.M$) or the base of the segment ($PSW.B$). The behavior of such an instruction is directly dependent on the actual values of the processor state, which are subject to alteration through the virtualization process. Such instructions must be privileged to ensure that the correct values are returned to the program.

The actual proof, which establishes the correspondence between the architecture and the existence of a VMM, requires some formalism. Rather than focusing on the formalism, let's first look at some counter-examples from the early era.

- Clearly, a single unprivileged, control-sensitive instruction would be a considerable concern. It would prevent the VMM from guaranteeing correct execution semantics, and possibly create security holes. In the formal model, this would include unprivileged updates to the segment register, even when M is unmodified. This was the key issue with the PDP-10 that motivated the paper, caused by the semantics of the `JRST 1` "return to user" instruction issued in user mode.

- Instructions that read the system state are behavior-sensitive and their use violates the equivalence criteria. If they are not privileged, the operating system may perform incorrectly as the hardware does not behave as defined in the architecture. Consider the simple case where an instruction would copy the value of `PSW.M` into a general-purpose register: the guest operating system could conclude, to much confusion, that it is running in user mode(!). The Intel x86-32 architecture has such instructions.

- Instructions that bypass virtual memory are behavior-sensitive, since their behavior depends on the actual value of the relocation registers. For example, some IBM VM/360 mainframes had such instructions. As long as the instruction is also privileged, it can be emulated. And provided the frequency of occurrence of that instruction is rare, this does not pose a performance problem.

The proof-by-construction highlights the fact that a VMM is really nothing more than an operating system. Both VMM and conventional operating systems share the fundamental construct of letting the untrusted component run directly on the hardware. Both must ensure that they remain in control and must correctly configure the hardware to achieve their goals. The difference is that one runs applications whereas the other runs entire virtual machines, with guest operating systems and applications. The key consequence of this difference is in the expected level of architectural support: it's a given for operating systems on nearly any computer architecture today, but much less obvious for VMMs.

2.3 RECURSIVE VIRTUALIZATION AND HYBRID VIRTUAL MACHINES

The paper includes two additional theorems, which cover requirements for recursive virtualization and hybrid virtual machines, respectively.

The formalism also extends to reason about **recursive virtualization**. Indeed, the idealized architecture that meets the criteria of the theorem can also support recursive virtual machines. In this scenario, a virtual machine can run a VMM in guest-supervisor mode rather than an operating system. That VMM of course can run multiple virtual machines, their guest operating systems and their respective applications. The second theorem of the paper formalizes the minimal set of additional assumptions posed on the VMM (but not the architecture) to support recursive virtual machines. In recursive virtualization, the VMM itself must run (unmodified) within a virtual machine. This topic of recursive virtualization has received renewed research interest in the last few years [33, 191].

The paper also introduces the notion of **hybrid virtual machines**, which discusses the case when the criteria fails but only according to specific circumstances. Consider again the PDP-10 JRST 1 instruction, which can be used both to return to user mode from supervisor mode, but equally to return from a subroutine when already in user mode. Clearly, this is catastrophic since the virtual machine would transition between the guest operating system and the application without a necessary trap into the VMM. Should that happen, the virtual machine would be executing application code when the VMM thinks that it is still in guest-supervisor mode.

Even though this is evidently unacceptable, the key observation for this particular instruction is that it is control-sensitive only when the virtual machine is in supervisor mode. When the virtual machine is in user mode, the specific instruction is not sensitive to virtualization.

Formally, let's define an instruction to be **user-sensitive** if it is behavior-sensitive or control-sensitive in supervisor mode, but **not** in user mode. The theorem states the following.

> **Theorem 3 [143]:** A hybrid virtual machine monitor may be constructed for any conventional third-generation machine in which the set of user-sensitive instructions are a subset of the set of privileged instructions.

Theorem 3 is much weaker than Theorem 1, and the resulting hybrid VMM is quite different than the traditional VMM. Specifically, a hybrid virtual machine monitor works around such limitations.

- The hybrid VMM acts like a normal VMM when the virtual machine is running applications, i.e., is in guest-user mode.

- Instead of direct execution, the hybrid VMM **interprets** in software 100% of the instructions in guest-supervisor mode. In other words, it interprets the execution of all paths through the guest-operating system. Despite the high overheads of interpretation, this approach may be practical as long as the portion of time spent in the operating system is small for the relevant workloads.

This third theorem precociously identified a key nuance that would play a crucial role much later with the emergence of paravirtualization and of hybrid solutions that combine direct execution with binary translation. Those will be discussed in Chapter 3.

2.4   DISCUSSION: REPLACING SEGMENTATION WITH PAGING

The original 1974 paper recognizes that virtual memory is a fundamental pre-requisite for any kind of virtualization support.

For the sake of simplicity, the authors assumed that virtual memory is implemented via a single relocation-bound register pair. As an exercise, consider a more realistic model where virtual memory is implemented via paging rather than segmentation.

Does the theorem still apply if virtual memory is implemented using paging? The short answer is yes, but a deeper analysis identifies subtleties not present in the third-generation model and these will play a practical role when building actual systems. In particular, instructions that access memory may be location-sensitive as the VMM must also be present somewhere in memory. Also, the guest operating system must remain protected from the application despite the fact that it is running in a de-privileged manner. These two issues will be revisited for the x86 architecture in §4.1 as the *address space compression* and *ring compression* challenges, respectively.

Finally, the VMM must compose the virtual address space by combining the page table mappings specified by the guest operating system with the mappings between guest-physical memory and host-physical memory. Whereas this is a simple matter of composition in an idealized segmented architecture, it is much more complex in page-based system. Indeed, there are two commonly used approaches to this problem, which are referred to as shadow page tables (see §3.2.5) and extended page tables (see §5.1).

2.5   WELL-KNOWN VIOLATIONS

In their article, Popek and Goldberg use the DEC PDP-10 as the example of an architecture that fails to meet the virtualization requirement. May the PDP-10 rest in peace. Instead, we list here some of the known violations of the reasonably modern era. The list of violations is not exhaustive. For the architectures discussed, we focus on the violations that had a practical impact in limiting the deployment of processor architectures with no claims of completeness. We do identify a few patterns of violations:

- **direct access to physical memory**: some architectural decisions expose directly physical memory into hardcoded portions of virtual memory. This occurs in particular with the MIPS architecture (see §2.5.1);

- **location-sensitive instructions**: other decisions expose the location in memory of certain sensitive data structures via unprivileged instructions (e.g., the global descriptor tables in x86, see §2.5.2); and

- **control-sensitive violations**: this occurs when the ISA clearly defines different semantics for the same instruction, depending on the privilege level (e.g., `iret`, `popf` on x86, see §2.5.2).

2.5.1   MIPS

The **MIPS** architecture is a classic RISC ISA. It has three execution modes: **kernel mode** (most privileged), **supervisor mode**, and **user mode** (least privileged). Only kernel mode can execute privileged instructions. The supervisor mode is really an alternate form of user mode, with the explicit benefit that it can access portions of virtual memory not available to the regular user mode.

First, the good news: This is a very promising start for an efficient hypervisor design. Indeed, the hypervisor would run in kernel mode, and could simply run the guest operating system in supervisor mode, emulating all privileged instructions according to their original semantic. The availability of supervisor mode leads to an important optimization: since supervisor virtual memory is protected from applications, the hypervisor can simply emulate transitions between guest operating system and applications as transitions between supervisor mode and user mode. Such transitions do not require changes to the configuration of virtual memory, or flushes of the TLB.

In the MIPS architecture, virtual memory is divided into multiple, fixed-sized regions, each with hardcoded attributes that determine which execution mode is required to access the region, how to relocate the address by either using the TLB (mapped) or via a hardcoded mask (unmapped), and whether the access should go through the cache hierarchy or not. Table 2.1 shows the regions for the 32-bit architecture. The 64-bit extensions create additional fixed-sized regions with the same general philosophy.

Table 2.1: Regions in the MIPS 32-bit architecture

Region	Base	Length	Access K,S,U	MMU	Cache
USEG	0x0000 0000	2 GB	✓,✓ ✓	mapped	cached
KSEG0	0x8000 0000	512 MB	✓,x,x	unmapped	cached
KSEG1	0xA000 0000	512 MB	✓,x,x	unmapped	uncached
KSSEG	0xC000 0000	512 MB	✓,✓,x	mapped	cached
KSEG3	0xE000 0000	512 MB	✓,x,x	mapped	cached

And now the bad news: The MIPS architecture is not virtualizable, first and foremost because of its use of regions. Indeed, the use of the regions is **location-sensitive** (a form of behavior-sensitivity), since it is a function of the privileged level of the processor. This has severe consequences, as any operating system expected to run in kernel mode will be compiled to use the KSEG0 and KSEG1 segments. Should a hypervisor attempt to run that OS in supervisor mode or user mode, every memory load/store instruction would cause a trap, thereby violating the efficiency criteria of virtualization. Since operating systems are generally not compiled or linked to run as position-independent code, the virtualization of the MIPS architecture requires at least the full recompilation of the guest operating system kernel. We discuss in §3.1 the architecture and compromise required in the implementation of Disco, a MIPS-based hypervisor.

2.5.2   X86-32

The Intel **x86-32** architecture is a notoriously complex CISC architecture, in part as it includes legacy support for multiple decades of backward compatibility. Over the years, the architecture introduced four main modes of operations (**real mode, protected mode, v8086 mode**, and **system management mode**), each of which enabled in different ways the hardware's segmentation model, paging mechanisms, four protection rings, and security features (such as call gates).

The x86-32 architecture was not virtualizable. It contained virtualization-sensitive, unprivileged instructions which violated the Popek and Goldberg criteria for strict virtualization [143]. This ruled out the traditional trap-and-emulate approach to virtualization.

Specifically, Robin and Irvine identified 17 problematic instructions that are sensitive and yet unprivileged [151]. Table 2.2 groups these instructions into 5 categories: instructions that manipulate the interrupt flag, manipulate segments registers and segment descriptors, can peek into the location of system data structures, and finally call gate-related instructions. The implications are severe and were well known. Indeed, before the introduction of VMware, engineers from Intel Corporation were convinced their processors could not be virtualized in any practical sense [74].

Table 2.2: List of sensitive, unprivileged x86 instructions

Group	Instructions
Access to interrupt flag	`pushf, popf, iret`
Visibility into segment descriptors	`lar, verr, verw, lsl`
Segment manipulation instructions	`pop <seg>, push <seg>, mov <seg>`
Read-only access to privileged state	`sgdt, sldt, sidt, smsw`
Interrupt and gate instructions	`fcall, longjump, retfar, str, int <n>`

The x86-32 architecture did provide one execution mode that was strictly virtualizable according to the Popek/Goldberg theorem: **v8086 mode** was specifically designed to run a 16-bit virtual machine running the 16-bit **Intel 8086** ISA. This mode was central to Windows 95/98 as it allowed these 32-bit operating systems to run legacy MS-DOS programs. Unfortunately, that mode was only capable of executing 16-bit virtual machines.

2.5.3   ARM

The **ARM** architecture is a RISC ISA. At a high-level from a virtualization perspective, ARM can be viewed as having two main execution modes, one or more **privileged modes** and **user mode**. Only privileged modes can execute privileged instructions. For example, ARMv6 has 7 processor modes, user mode and 6 privileged modes,[1] while ARMv8 has effectively user mode and a single

---

[1]See pages A2-3 to A2-5 in the *ARM Architecture Reference Manual* [20] for more information.

privileged mode, although ARMv8 terminology refers to them as **exception levels**.[2] Each mode has a number of **banked registers**, which means, for instance, register 13 points to a different physical register in each mode. The differences between the privileged modes only concern the banked registers and can be ignored for the purposes of this discussion.

The ARM architecture was not virtualizable. It contained virtualization-sensitive, unprivileged instructions, which violated the Popek and Goldberg criteria for strict virtualization [143]. This ruled out the traditional trap-and-emulate approach to virtualization. Dall and Nieh identified 24 problematic instructions that are sensitive and yet unprivileged [59]. The specific instructions were identified for ARMv6, but are also present in other versions of the architecture. For example, ARMv7 is quite similar to ARMv6 in this respect. There are similar problematic instructions even in the most recent version of the ARM architecture, ARMv8. The instructions deal with user mode registers, status registers, and memory accesses that depend on CPU mode, as listed in Table 2.3.

Table 2.3: ARM sensitive instructions

Description	Instructions
User mode	LDM (2), STM (2)
Status registers	CPS, MRS, MSR, RFE, SRS, LDM (3)
Data processing	ADCS, ADDS, ANDS, BICS, EORS, MOVS, MVNS, ORRS, RSBS, RSCS, SBCS, SUBS
Memory access	LDRBT, LDRT, STRBT, STRT

There are various load/store multiple instructions that access user mode registers when in privileged mode. However, these instructions are defined by the ARM architecture as **unpredictable** when executed in user mode, which means the result of the instruction cannot be relied upon. For example, one common implementation of an unpredictable instruction is that it is ignored by the processor and does not trap.

Status register instructions relate to special ARM status registers, the **Current Program Status Register** (CPSR) and **Saved Program Status Register** (SPSR). CPSR specifies the current mode of the CPU and other state information, some of which is privileged. SPSR is a banked register available in a number of the privileged modes. The basic ARM protection mechanism works by copying the CPSR to the SPSR when the processor enters the respective privileged mode via an exception so that CPU state information at the time of the exception can be determined. Similarly, the SPSR is copied to the CPSR at various times to update the mode of the CPU as the preferred exception return mechanism. There are three types of problems arising with these instructions. First, MRS can be used in any mode to read the CPSR and determine

---

[2]Although in ARMv8 terminology, what we refer to here as a mode is more precisely called an exception level, we will continue to use the term mode even when referring to ARMv8 to avoid architecture version-specific terminology and provide a more consistent comparison with other architectures.

the CPU mode, although this is no longer the case for ARMv8. Second, CPS is used to write the CPSR and change the CPU mode, but is ignored when executed in user mode. Third, other status register instructions are unpredictable when executed in user mode. For example, LDM (3) copies the SPSR into the CPSR as part of the instruction when executed in privileged mode, but is unpredictable when executed in user mode.

Almost all data processing instructions have a special version, which replaces the CPSR with the SPSR in addition to their ALU operation. These instructions are denoted by appending an S to the normal instruction. They are designed for exception return so that, for example, it is possible to jump to userspace from a kernel and change modes at the same time. These instructions are unpredictable when executed in user mode.

Memory access instructions access memory using a different CPU mode from the one being used for execution. The virtual memory system on ARM processors uses access permissions to limit access to memory depending on the CPU mode. The architecture defines four instructions called load/store with translation that access memory using user mode access permissions even though the CPU is in a privileged mode, which will therefore trap due to memory access violations. However, when executed in user mode, these instructions behave as regular memory access instructions. For example, when running the OS kernel in user mode as part of a VM, the semantics of the instructions are different than when running the OS in kernel mode, so the OS may not see the memory access faults it expects.

2.6   FURTHER READING

The proof in the Popek and Goldberg's paper was done for an idealized uniprocessor, third-generation architecture with a single virtual memory segment. Of course, current-generation ISAs are much more complex: they support multiprocessor architectures with cache-coherency and well-defined, subtle, memory consistency models. They support virtual memory via paging, either as an alternative to segmentation or in addition to segmentation. To make matters worse, the devil is in the details, in particular on CISC architectures with their numerous legacy execution modes and instructions.

With those caveats, the proof-by-construction provides a framework to reason about virtualization. The classification of control-sensitive and behavior-sensitive instructions is a helpful guide to evaluate any modern computer architecture. Motivated readers are encouraged to read the seminal paper from Popek and Goldberg in enough detail so that they can reason about known architectures [143].

The limitations of x86-32 were documented by Robin and Irvine [151]; the paper, however, incorrectly concludes that no practical and secure virtualization solution can be built as a result. Chapter 4 describes how Intel later relied on Popek and Goldberg's theorem as a framework to add virtualization support in x86-64 processors [171].

The limitations of ARMv6 and earlier ARM architectures were documented by Dall and Nieh [59]. Chapter 7 describes how ARM later added virtualization support in ARMv7 processors [39] and how the architecture relates to Popek and Goldberg's theorem.

CHAPTER 3

Virtualization without Architectural Support

This chapter is about the past. Practitioners who are only interested in understanding how virtualization operates on contemporary hardware and hypervisors may be tempted to skip to Chapter 4.

However, the past does remain relevant to any computer scientist who needs to understand how we got to today's situation, and to those interested in the specific techniques developed in an earlier era. Indeed, many of the software techniques precedently developed have applications outside of virtualization. In addition, computer architects reading this chapter will get an appreciation of the unanticipated consequences of architectural decisions made both before and after the introduction of hardware support for virtualization. Case in point, both VMware and Xen relied on segmentation for protection on 32-bit architectures, as we will explain in this chapter. Segmentation was removed with the introduction of VT-x on x86-64 as it was no longer needed to build hypervisors. However, segmentation had other applications, e.g., to provide protection to **lightweight virtual machines** such as VX32 [71] and the original Google Native Client [190].

The violations of the Popek and Goldberg theorem for MIPS, x86-32, and ARM were previously described in §2.5. As a direct consequence of these violations, no hypervisor—at least no hypervisor built using the techniques anticipated by Popek and Golberg of **trap-and-emulate** combined with **direct execution**—can be built to simultaneously address the equivalence, safety, and performance requirements. This chapter describes how three systems—Disco, VMware Workstation, and Xen—each use different techniques or made different tradeoffs between equivalence, safety, and performance to work around the limitations of the theorem.

Table 3.1 provides a map to the rest of this chapter. Each section presents, as a case study, the salient features of a particular hypervisor, and the sections are ordered chronologically with respect to the introduction of each system.

3.1 DISCO

Disco [44, 81] was a research hypervisor designed for the **Stanford FLASH multiprocessor** [117] built using MIPS [85] processors. Disco is credited for its historical role in the resurgence of virtual machines [2].

The primary motivation for Disco was the inability for operating systems to readily take advantage of new hardware trends. In the case of FLASH, the innovation was scalable shared-memory multiprocessors. Although research groups had demonstrated that prototype operating

Table 3.1: Case studies of hypervisors designed for architectures with no virtualization support

	Disco	VMware Workstation	Xen	KVM for ARM
Architecture	MIPS	x86-32	x86-32	ARMv5
Hyp type	Type-1	Type-2 (§4.2.4)	Type-1 with dom0 (§4.4)	Type-2 (§4.5)
Equivalence	Requires modified kernel	Binary-compatible with selected kernels	Required modified (paravirtualized) kernels (§4.3)	Required modified (lightweight paravirtualized kernels (§4.5)
Safety	Via de-privileged execution using strictly virtualized resources	Via dynamic binary translation; isolation achieved via segment truncation	Via de-privileged execution with safe access to physical names	Via de-privileged execution using strictly virtualized resources
Performance	Via localized kernel changes and L2TLB (§4.1.2)	By combining direct execution (or applicastions) with adaptive dynamic binary translation (§4.2.3)	Via paravirtualization of CPU and IO interactions	Via paravirtualization of CPU and IO interactions

systems could scale and address fault-containment challenges [47, 172], these designs required significant OS changes. In particular, scalability was achieved by the partitioning of the system into scalable units that communicated with each other like a distributed system. In addition, machines such as FLASH had a non-uniform memory architecture, which required additional complex changes within the virtual memory subsystem of the operating system [175]. So, even though prototypes could be built, the complexity of the approach made their relevance questionable.

Disco addressed this same problem very differently, by developing a new twist on the relatively old idea of hypervisors and virtual machines. Rather than attempting to modify existing operating systems to run on scalable shared-memory multiprocessors, Disco inserted an additional layer of software between the hardware and the operating system. In this design, the hypervisor handled the changes required to support scalable shared-memory computing, in particular resource management and transparent resource sharing (described in the original paper) and fault containment, described in Cellular Disco [81].

Disco was architected around the traditional trap-and-emulate approach formalized by Popek and Goldberg. Disco itself runs in kernel mode; guest operating systems (which were designed to run in kernel model) run in supervisor mode, and applications run in user mode. Disco handled the equivalence, safety, and performance requirements as follows:

Equivalence: Disco did not attempt to run binary-compatible kernels as guest operating systems. MIPS instructions are location-sensitive as a consequence of the hardcoded virtual memory segment ranges. Unfortunately, many MIPS operating systems, including IRIX 5.3, place the kernel code and data in the KSEG0 segment, which is not accessible in supervisor mode (see §2.5.1). As a result, Disco requires that the operating system be relocated from the unmapped segment of the virtual machines (KSEG0) and into a portion of the mapped supervisor segment of the MIPS processor (KSSEG). Making these changes to IRIX required modifying two header files that describe the virtual address space layout, changing the linking options, as well as 15 assembly statements. Unfortunately, this meant that IRIX had to be entirely re-compiled to run on Disco.

Safety: Disco isolated the VM without making assumptions about the guest. It relied on virtual memory and the MIPS' three privilege levels (kernel, supervisor, user) to protect the hypervisor from the VM, and the VM's guest operating system from its applications. Specifically, supervisor mode code (i.e., the guest operating system) cannot issue privileged instructions, which trigger traps. Disco virtualized all resources and hid the physical names from the VM, e.g., the VM only operated with guest-physical memory without any visibility into host-physical mappings. Disco's approach to memory management is discussed in §3.1.3.

Performance: In their idealized model, Popek and Goldberg assumed that traps were rare, and therefore that the performance requirement can be met as long as the guest instructions run directly on the processor and the hypervisor handles the traps efficiently. This assumption does not hold on RISC architectures such as MIPS. To address this, Disco introduced special memory pages as an alternative to read-only privileged registers, hypercalls (see §3.1.1) and a larger TLB (see §3.1.2).

3.1.1   HYPERCALLS

In a strict approach to virtualization, the guest operating system runs identically in a virtual machine as it would on actual hardware. Disco, because of the limitations of MIPS, requires that its guest kernels be recompiled anyway. Disco relied on special memory pages to eliminate having to take a trap on every privileged register access, as this can cause significant overheads when running kernel code such as synchronization routines and trap handlers that frequently access privileged registers. To reduce this overhead, Disco augments the instruction set architecture through special pages in the address space containing frequently accessed privileged registers. The guest operating system is then patched in a handful of locations to convert certain privileged instructions that access these registers into regular load instructions.

Disco further deviates from the strict approach through its use of hypercalls. A **hypercall** is a higher-level command issued by the guest operating system to the hypervisor. It is analogous to a system call, which is issued by the application to the operating system. Hypercalls are used to provide higher-level information, in particular around the use of memory resources. For example, Disco has a hypercall to free a guest-physical page. This is called by the guest whenever a page is put on the operating system's free list. To the virtual machine, the only side effect is that the content of that page will be zeroed out upon next use. To the global system, however, this provides the ability to substantially reduce the global host-physical memory footprint of the consolidated workload.

3.1.2   THE L2TLB

Popek and Golberg assumed that traps resulting from the de-privileged execution of the guest operating systems would be sufficiently rare that the hypervisor's trap-and-emulate approach would not cause significant performance problems. This was a reasonable assumption in mainframe-era computers characterized by complex privileged instructions issued relatively rarely during operating system execution.

The Popek/Goldberg further model assumed that virtual memory was implemented through segmentation. MIPS implements virtual memory through paging, with TLB misses handled in software by a **TLB miss handler**. Processors of that era had small TLBs (e.g., with ≤ 64 entries), leading to frequent misses. As a result, the TLB itself was virtualized with low overheads to meet the performance requirement of virtualization.

Software-reloaded TLBs introduce a few complications. First, the page table format is defined by the operating system, and not the architecture. An OS-agnostic hypervisor would therefore be incapable of handling a TLB miss directly. Instead, the hypervisor has to transfer execution to the guest's TLB miss handler. Unfortunately, that sequence typically contains multiple privileged instructions, causing high virtualization overheads. Second, a workload executing on top of Disco will suffer an increased number of TLB misses since the TLB is additionally used for all operating system references (now using KSSEG rather than KSEG0). Third, virtualization introduces an additional level of indirection between guest-physical and host-physical memory, with each virtual machine having its own guest-physical address space.

To lessen the performance impact, Disco cached recent virtual-to-host physical translations in a second-level software TLB (or L2TLB). On each TLB miss, Disco's TLB miss handler first consults the second-level TLB. If it finds a matching virtual address it can simply place the cached mapping in the hardware TLB, otherwise it forwards the TLB miss exception to the operating system running on the virtual machine. The effect of this optimization is that virtual machines appear to have much larger TLBs than the MIPS processors.

Each virtual CPU has its own L2TLB, which replaces the original TLB in the architecture of the virtual machine. The only visible change to the architecture is the number of entries in the TLB itself. Fortunately, the impact on operating systems is very contained. In fact, the only complication from such a large TLB was the need for the operating system to recycle address space identifiers associated with processes, the details of which can be found in [44].

3.1.3   VIRTUALIZING PHYSICAL MEMORY

The primary research motivation of Disco was to provide efficient resource management on top of a scalable shared-memory multiprocessor. One goal was to ensure that a distributed workload running on a cluster of virtual machines would have comparable memory usage patterns as the same workload running as a collection of processes on a non-virtualized deployment. Fortunately, the additional level of indirection between guest-physical and host-physical memory provides the enabling mechanism to solve this problem *as long as* host-physical names are *not* visible to the VM. Then, the mapping between guest-physical and host-physical can take many forms and change over time without impacting the execution of the VM. In the case of Disco, mappings could change as the result of either (1) transparent page migration and replication, which improve CC-NUMA locality or (2) transparent memory sharing, which reduced overall memory pressure and disk I/O bandwidth. In later commercial systems, mappings would change as the result of memory ballooning and content-based page sharing [177].

In its most basic form, each host-physical page (of the machine) is used by at most one guest-physical page of a single virtual machine. However, Disco's memory management mechanisms enabled three additional use cases:

Many-to-one mappings: Disco implemented a copy-on-write mechanism that operates within and across virtual machines. As in the well-known copy-on-write mechanism of operating systems, corresponding entries are inserted read-only into the TLB. Upon a page fault, due to a legitimate write by the virtual machine (application or operating system), the hypervisor simply copies the content to a newly allocated host-physical page. The mechanism was exposed transparently to the virtual machines through I/O operations: disk reads from the same sector and network traffic between two virtual machines on the same server could be shared transparently. In addition, the mechanism was also exposed explicitly through a remap function with the semantics of bcopy routine that used a hypercall to remap the page whenever possible.

One-to-many mappings: Disco provided transparent page replication for read-only pages that suffer frequent cache misses. On a cc-NUMA machine such as FLASH, this optimization ensures that such pages always suffer cache misses served by the local NUMA node rather than a remote node. In Disco's implementation, two virtual CPU of the same VM can therefore map the same guest-physical page to two distinct host-physical pages.

NUMA migration: Unfortunately, page replication is only effective for read-only pages. When a page is accessed (read and written) by a single NUMA node, Disco transparently migrates the page to that node. To do so, it must allocate a new host-physical page and then atomically copy its content, change the guest-physical to host-physical mapping and invalidate the hardware TLB and L2TLB.

These three mechanisms each require subtle interactions between the physical memory management aspect, which manages the mappings between guest-physical and host-physical memory, and virtual memory and TLB miss handling logic. Disco's core data structures were specifically designed with these in mind. For example, Disco maintains an inclusion property within the TLB hierarchy to efficiently determine when hardware invalidations are required. Here also, the details are in [44].

3.2   VMWARE WORKSTATION—FULL VIRTUALIZATION ON X86-32

VMware Workstation 1.0 was released in 1999 and was the first VMM for the 32-bit x86 architecture [45, 46, 65, 162]. As a commercial product, the company's vision was a virtualization layer useful on commodity platforms built from x86-32 CPUs and primarily running the Microsoft Windows operating systems (a.k.a. the *WinTel* platform). The benefits of virtualization could help address some of the known limitations of the WinTel platform, such as application interoperability, operating system migration, reliability, and security. In addition, virtualization could easily enable the co-existence of operating system alternatives, in particular Linux.

With its focus on commercial, closed-source operating systems, the x86 computing environment was sufficiently different to require new approaches to virtualization; recompiling the kernel, as was previously done in Disco or later in Xen, was not an option. Furthermore, the x86 industry had a disaggregated structure. Different companies independently developed x86 processors, computers, operating systems, and applications. System integrators then combine these components into a supported "out-of-the-box" solution. For the x86 platform, virtualization would need to be inserted without changing either the existing hardware or the existing software of the platform. As a result, VMware Workstation was architected as:

- a pure virtualization solution, compatible with existing, unmodified guest operating systems; and

- a type-2 hypervisor for existing Linux and Windows host operating systems.

VMware Workstation adapted Popek and Goldberg's three core attributes of a virtual machine to x86-based target platform as follows:

Equivalence: As discussed in §2.5.2, the x86-32 architecture has 17 virtualization-sensitive, non-privileged instructions, which violated the Popek and Golberg criteria for strict virtualization [151]. This ruled out the traditional trap-and-emulate approach to virtualization. Indeed, engineers from Intel Corporation were convinced their processors could not be virtualized in any practical sense [74]. VMware's solution combines direct execution (used whenever possible and in particular to run applications) with dynamic binary translation (whenever required, and in particular when running guest operating systems). Dynamic binary translation is an efficient form of emulation; see §3.2.2.

Safety: A hypervisor must guarantee the isolation of the virtual machine without making any assumptions about the software running inside. VMware Workstation configured the hardware, and in particular made extensive use of **segment truncation** to isolate the virtual machines (see §3.2.3). To simplify testing and reduce the likelihood of introducing security vulnerabilities in untested code paths, VMware Workstation only supported the subset of the x86-32 architecture necessary to run some specified, supported guest operating systems. Any unsupported requests, e.g., attempting to execute code at %cpl[1]=1 or %cpl=2, which never happens with any supported guest operating system, would simply abort execution.

Performance: As a design goal, VMware aimed to run relevant workloads at near native speeds, and in the worst case to run them on then-current processors with the same performance as if they were running on the immediately prior generation of processors without virtualization. Such performance levels would allow users to use virtual machines as effectively as real machines. This was based on the observation that most x86 software wasn't designed to only run on the latest generation of CPUs.

To achieve such performance goals, VMware Workstation required total control over the hardware during execution, despite its type-2 architecture and the presence of a host operating system. §3.2.3 describes how the segment table and the CPU are configured to reduce the overheads of the dynamic binary translator. §3.2.4 describes the modular design of the type-2 hypervisor into three distinct components: a user-level process of the host operating system, a kernel module within the host operating system and the **VMware VMM**, which runs in total control of the hardware. This is enabled by the **world switch**, which gives the VMware VMM total control in co-location with the host operating system. Finally, §3.2.5 describes shadow paging, which allows the virtual machine (including the guest operating system) to directly use the MMU of the underlying hardware.

3.2.1   X86-32 FUNDAMENTALS

A short introduction to the x86-32 architecture is required before explaining how VMware virtualized it. Readers already familiar with it can skip to the solution in §3.2.2. For the CPU, the architecture specifies legacy and native execution mode, four privilege levels, and an additional I/O privilege levels.

- The processor has a native execution mode, called `protected` mode. It contains three additional non-native mode which are the legacy `real`, `system management`, and `v8086` modes.

- In protected mode, the current privilege level (or `cpl`) separates kernel execution (%cpl=0) from user-level execution (%cpl>0). There are four modes although in practice user-level code runs at %cpl=3 and the intermediate modes are unused.

- The I/O privilege level (`iopl`) further allows user-level code to enable or disable interrupts

---

[1]cpl=current privilege level in x86 protected mode, a.k.a. the *ring*

As far as memory, the architecture includes both segmentation and paging. Segmentation is handled by six segment registers which correspond to code (%cs), stack (%ss), data (%ds), and extra segments (%es, %fs, %gs). Each segment defines through a base and a limit a portion of the 32-bit linear address. Paging then converts the linear address into a physical address. The x86-32 architecture specifies a three-level page table structure rooted at privilege register %cr3. On a TLB miss, the processor directly walks the page table structure and inserts the mapping in the TLB without relying on a software handler.

3.2.2   VIRTUALIZING THE X86-32 CPU

A hypervisor built for a virtualizable architecture uses the trap-and-emulate approach described by Popek and Goldberg. This technique is also known as **direct execution** (DE) as the VM instructions directly execute on the processor. With a design goal to run unmodified guest operating systems on a non-virtualizable architecture, direct execution alone is not a strategy.

An alternative would have been to employ an all emulation approach. The experience with the SimOS [153] machine simulator showed that the use of techniques such as **dynamic binary translation** (DBT) running in a user-level program could limit overheads of complete emulation to a 5× slowdown. While that was fast for a machine simulation environment, it was clearly inadequate for our performance requirements.

VMware's solution to this problem combined two key insights. First, although direct execution could not be used to virtualize the entire x86 architecture, it could actually be employed most of the time, in particular to run applications. The second key insight was that by properly configuring the hardware, particularly by using the x86 segment protection mechanisms carefully, system code under dynamic binary translation could also run at near-native speeds; this will be discussed below in §3.2.3.

Figure 3.1 shows the actual algorithm dynamically determining whether direct execution can be used or whether dynamic binary translation must be used [65]. The algorithm depends on the state of a few specific registers of the virtual machine and leads to the following decisions:

- (line #1–#3): %cr0.pe is set only in protected mode and v8086 mode. DBT is required when the VM is in either `real` mode or `system management` mode. For example, the BIOS runs in real mode;

- (line #4–#6): The x86 v8086 mode can always be used directly to virtualize guests in v8086 mode using DE, e.g., to run MS-DOS emulation in Windows 95 [111];

- (line #7–#9): DBT is required whenever the VM can control the interrupt flag. The disjunction includes the common case when the guest executes in kernel mode execution (cpl=0), but also the rare case where a user-level process can establish critical regions by disabling interrupts (e.g., on Linux using the `iopl(2)` system call);

Input: Current state of the virtual CPU

Output: True if the direct execution subsystem may be used;
             False if binary translation must be used instead

1: **if** $!cr0.pe$ **then**
2:     return false;
3: **end if**
4: **if** $eflags.v8086$ **then**
5:     return true
6: **end if**
7: **if** $(eflags.iopl \geq cpl) || (!eflags.if)$ **then**
8:     return false;
9: **end if**
10: **for all** $seg \leftarrow (cs, ds, ss, es, fs, gs)$ **do**
11:     **if** "seg is not shadowed" **then**
12:         return false;
13:     **end if**
14: **end for**
15: return true

Figure 3.1: x86 virtualization engine selection algorithm.

- (line #10–14): DBT is required whenever the hidden content any of the six segment descriptor registers is not recoverable from memory. This is a somewhat obscure corner-case that must be handled to correctly run Windows 95; and

- DE can be used in the remaining cases, and in particular at user level when interrupts cannot be disabled.

If we exclude the corner cases, the algorithm has the VMware VMM use DBT to execute the guest operating system and DE to run the applications. This is good news since the amount of kernel code is bounded, and most applications spend the majority of the execution time at user level.

This algorithm has two interesting properties: (i) it does not make any assumption on the guest instructions that may execute in the future but instead executes in $O(1)$; and (ii) it can be implemented in a handful of assembly instructions through a careful encoding of the state of the virtual processor.

Formal considerations: Popek and Goldberg discuss a similar situation with their **hybrid virtual machines** (see §2.3 and the 3rd theorem [143]), which applies to architectures in which all **user-sensitive instructions** are privileged. The x86-32 architecture *nearly* falls into that category:

sgdt, sidt, sldt, and smsw are the only non-privileged, user-sensitive instructions. Fortunately, even Intel's manual describes them as *available but not useful to applications* [102]. In addition, segment truncation, which is used to isolate the VMware VMM, is visible to application via the lsl instruction. VMware Workstation's use of direct execution violates the equivalence requirement only, with few practical consequences other than to provide an easy way to determine if an application is running in a virtualized environment.

3.2.3   THE VMWARE VMM AND ITS BINARY TRANSLATOR

The main function of the VMM was to virtualize the CPU and the main memory. At its core, the VMware VMM combined a direct execution subsystem with a dynamic binary translator. In simple terms, direct execution was used to run the guest applications and the dynamic binary translator was used to run the guest operating systems.

Dynamic binary translation (DBT) [84] is an efficient form of emulation. Rather than interpreting the instructions one by one, DBT compiles a group of instructions, often a basic block, into a fragment of executable code. The code is stored in a large buffer called the translation cache so that it can be reused multiple times. DBT comes with some well-known optimizations such as chaining, which allows for direct jumps between compiled fragments [53]. VMware relies on DBT so that, instead of executing or interpreting the original virtual machine instructions, the processor executed natively the corresponding translated sequence within the translation cache.

The performance of DBT, and in particular of system-level DBT, is very sensitive to the configuration of the hardware. For example, Embra [184] suffers from a 5× slowdown largely due to the cost of MMU emulation. In general, DBT systems also runs in the same address space as the software they emulate. With VMware, the challenge was to ensure that the VMware VMM could share an address space safely with the virtual machine without being visible to the virtual machine, and to execute this with minimal performance overheads. Given that the x86 architecture supported both segmentation-based and paging-based protection mechanisms, a solution might have used either one or both mechanisms. For example, operating systems that use the **flat memory model** only rely on paging (and not segmentation) to protect themselves from applications.

The VMware VMM used segmentation, and segmentation only, for protection. The linear address space was statically divided into two regions, one for the virtual machine and one for the VMM. Virtual machine segments were *truncated* by the VMM to ensure that they did not overlap with the VMM itself.

Figure 3.2 illustrates VMware's use of **segment truncation**, with the example of a guest operating system that employs the flat memory model. Applications running at %cpl=3 ran with truncated segments, and were additionally restricted by their own operating systems from accessing the guest operating system region using page protection.

When running guest kernel code via binary translation, the hardware CPU was at %cpl=1. Binary translation introduced a new and specific challenge since translated code contained a mix

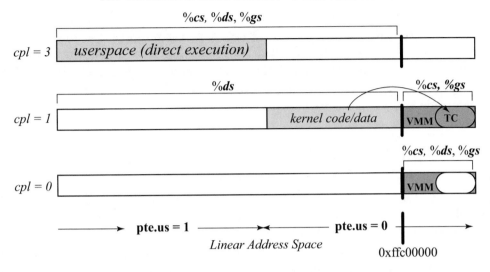

Figure 3.2: Using segment truncation to protect the VMware VMM [45]. In this example, the virtual machine's operating system is designed for the flat memory model. Applications run under direct execution at user-level (`cpl=3`). The guest operating system kernel runs under binary translation, out of the translation cache (TC), at `cpl=1`.

of instructions. Some needed to access either the VMM area (to access supporting VMM data structures), while others needed to access the virtual machine's portion of the linear address space.

The solution was to rely on hardware protection instead of any run-time memory checks. Specifically, the VMware VMM reserves one segment register, `%gs`, to always point to the VMM area: instructions generated by the translator used the `<gs>` prefix to access the VMM area, and the binary translator guaranteed (at translation time) that no virtual machine instructions would ever use the `<gs>` prefix directly. Instead, translated code used `%fs` for virtual machine instructions that originally had either an `<fs>` or `<gs>` prefix. The three remaining segments (`%ss`, `%ds`, `%es`) were available for direct use (in their truncated version) by virtual machine instructions.

Of course, virtual machine instructions may have had a legitimate, and even frequent, reason to use addresses near the top of the linear address space, where the VMM actually resides. As expected, segment truncation triggered a general protection fault for every such reference, which can be emulated by the VMM. To reduce the number of such traps, VMware Workstation relied on **adaptive binary translation** as an optimization to eliminate most general protection faults at run-time. Adaptive binary translation is premised on the notion that the same few locations in the guest kernel cause nearly all such general protection faults. It re-translates the basic block and replaces the original instruction (that causes the trap) with a specific sequence that safely emulates the memory access without causing the trap.

Figure 3.2 also illustrates the role of page-based protection (`pte.us`). Although not used to protect the VMM from the virtual machine, it is used to protect the guest operating system from its applications. The solution is straightforward: the `pte.us` flag in the actual page tables was the same as the one in the original guest page table. Guest application code, running at `%cpl=3`, was restricted by the hardware to access only pages with `pte.us=1`. Guest kernel code, running under binary translation at `%cpl=1`, did not have the restriction.

3.2.4   THE ROLE OF THE HOST OPERATING SYSTEM

VMware Workstation is a type-2 hypervisor that appears to run on top of a **host operating system** such as Linux or Windows, when in fact the hypervisor, and specifically the VMware VMM, has full control of the CPU when it executes the virtual machine [46, 162]. Figure 3.3 illustrates the hosted architecture of VMware Workstation, which consists of three components: (i) a user-level program (VMX), responsible for all interactions with the end-user and the host operating system, in particular for I/O emulation purposes; (ii) a small device driver installed within the kernel (VMM Driver); and (iii) the VMware VMM, which runs at the same level as the host operating system, but in a disjoint environment in which the host operating system has been temporarily suspended and removed from virtual memory.

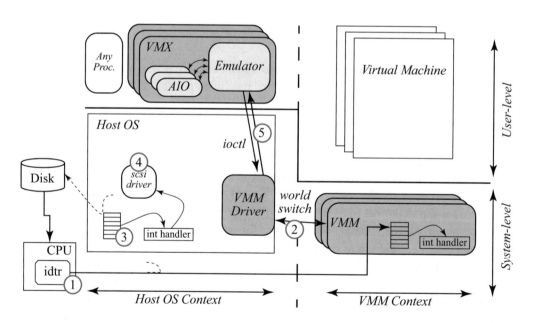

Figure 3.3: The VMware Hosted Architecture [45]. VMware Workstation consists of the three shaded components. The figure is split vertically between host operating system context and VMM context, and horizontally between system-level and user-level execution.

The **world switch** is the low-level mechanism that frees the VMware VMM from any interference from the host operating system, and vice-versa. Similar to a traditional context switch, which provides the low-level operating system mechanism that loads the context of a process, the world switch loads and restores the VMware VMM context, as well as the reverse mechanism that restores the host operating system context. The difference between a traditional context switch and the world switch is its scope: the context switch only concerns itself with callee-saved registers, the stack, and the address space and can furthermore assume that the kernel is identically mapped in the outgoing and incoming address spaces. In contrast, the world switch must save everything, including the x86-32 segment descriptor tables and the interrupt descriptor tables, and must assume disjoint address spaces.

Although subtle in a number of ways, the implementation was quite efficient and very robust. It relied on two basic observations: first, any change of the linear-to-physical address mapping via an assignment to %cr3 required that at least one page—the one containing the current instruction stream—had the same content in both the outgoing and incoming address spaces. Otherwise, the instruction immediately following the assignment of %cr3 would be left undetermined. The second observation is that certain system-level resources could be undefined as long as the processor did not access them, e.g., the interrupt descriptor table (as long as interrupts are disabled) and the segment descriptor tables (as long as there are no segment assignments). By undefined, we mean that their respective registers (%idtr and %gdtr) could point temporarily to a linear address that contained a random, or even an invalid page. With these two observations, VMware developed a carefully crafted instruction sequence that saved the outgoing context and restored an independent one. In an early version of VMware Workstation, the cross-page instruction sequence consisted of only 45 instructions executing symmetrically in both directions, with disabled interrupts.

Figure 3.3 puts the world switch into perspective by illustrating the sequence of steps that occur when an external interrupt (e.g., a disk interrupt) occurs while the VMware VMM is in total control of the hardware. This example highlights that fact that the host operating system is really oblivious to the existence of the VMware VMM.

1. The hardware interrupts the execution of the virtual machine. The interrupt descriptor table (%idtr) points to a VMware VMM interrupt handler.

2. As the VMware VMM cannot handle I/O requests directly, the interrupt handler initiates a world switch back to the host operating system context.

3. Within the host operating system, the VMM Driver implements the other end of the world switch. When the cause of the world switch is an external interrupt, the VMM Driver issues in software the same interrupt received in step #1.

4. At this point, the host operating system handles the interrupt as if it had occurred in the normal host operating system context, and will call the appropriate subsystems, in this case the disk handler.

5. After the interrupt has completed, execution resumes where it was interrupted, i.e., within the VMM Driver, which is understood by the host operating system as merely handling an `ioctl` system call. The ioctl completes and returns control back to the VMX application. This transition back to userspace allows the host operating system to make scheduling decisions, and the VMX application to process asynchronous I/O requests (AIO). To resume execution of the virtual machine, the VMX issues the next ioctl.

3.2.5   VIRTUALIZING MEMORY

This section describes how the VMware VMM virtualized the linear address space shared with the virtual machine, and how the VMM virtualized guest-physical memory and implemented a central mechanism called memory tracing.

The VMM was responsible for creating and managing the page table structure, which was rooted at the *hardware* `%cr3` register while executing in the VMM context. The challenge was in managing the hardware page table structure to reflect the composition of the page table mappings controlled by the guest operating system (linear to guest-physical) with the mappings controlled by the host operating system (guest-physical to host-physical) so that each resulting valid `pte` always pointed to a page previously locked in memory by the VMM Driver. This was a critical invariant to maintain at all times as to ensure the stability and correctness of the overall system. In addition, the VMM managed its own 4 MB address space.

Memory tracing: Memory tracing provided a generic mechanism that allowed any subsystem of the VMM to register a *trace* on any particular page in guest physical memory, and be notified of all subsequent writes (and in rare cases reads) to that page. The mechanism was used by VMM subsystems to virtualize the MMU and the segment descriptor tables, to guarantee translation cache coherency, to protect the BIOS ROM of the virtual machine, and to emulate memory-mapped I/O devices. When composing a `pte`, the VMM respected the trace settings so that pages with a write-only trace were always inserted as read-only mappings in the hardware page table. Since a trace could be requested at any point in time, the system used the backmap mechanism to downgrade existing mappings when a new trace was installed. As a result of the downgrade of privileges, a subsequent access by any instruction to a traced page would trigger a page fault. The VMM emulated that instruction and then notified the requesting module with the specific details of the access, such as the offset within the page and the old and new values.

Unfortunately, handling a page fault in software took close to 2000 cycles on the processors of the time, making this mechanism very expensive. Fortunately, nearly all traces were triggered by guest kernel code. Furthermore, there is an extreme degree of instruction locality in memory traces, and in particular only a handful of instructions of a kernel modified page table entries and triggered memory traces. For example, Windows 95 has only 22 such instruction locations. As with segment truncation, the DBT relied on adaptive binary translation to generate an alternative code sequence that avoided the page fault altogether.

This was not a mere optimization. As we will discuss later in §4.2.3, memory tracing is only practical when used in conjunction with adaptive binary translation [3].

Shadow page tables: The first application of the memory tracing mechanism was actually the MMU virtualization module itself, responsible for creating and maintaining the page table structures (pde, pte) used by the hardware.

Like other architectures, the x86 architecture explicitly calls out the absence of any coherency guarantees between the processor's hardware TLB and the page table tree. Rather, certain privileged instructions flush the TLB (e.g.,invlpg, mov %cr3). A naive virtual MMU implementation would discard the entire page table on a TLB flush, and lazily enter mappings as pages are accessed by the virtual machine. Unfortunately, this generates many more hardware page faults, which are in order of magnitude more expensive to service than a TLB miss.

So, instead, the VMM maintained a large set of shadow copies of the guest operating system's pde/pte pages, as shown in Figure 3.4. By putting a memory trace on the corresponding original pages (in guest-physical memory), the VMM was able to ensure the coherency between a very large number of guest pde/pte pages and their counterpart in the VMM. This use of shadow page tables dramatically increased the number of valid page table mappings available to the virtual machine at all times, even immediately after a context switch. In turn, this correspondingly reduced the number of spurious page faults caused by out-of-date page mappings. This category of page faults is generally referred to as **hidden page faults** since they are handled by the VMM and not visible to the guest operating system. The VMM could also decide to remove a memory trace (and of course the corresponding shadow page), e.g., when a heuristic determined that the page was likely no longer used by the guest operating system as part of any page table.

Figure 3.4 shows that shadowing is done on a page-by-page basis, rather than on an address space by address space basis: the same pte pages can be used in multiple address spaces. In fact, this is commonly used by all operating systems which map the kernel portion identically into all address spaces. When such sharing occurred in the operating system, the corresponding shadow page was also potentially shared in the shadow page table structures. The VMM shadowed multiple pde pages, each potentially the root of a virtual machine address space. So even though the x86 architecture does not have a concept of address-space identifiers, the virtualization layer emulated it.

The figure also illustrates the special case of the top 4 MB of the address space which is always defined by a distinct pte page. This portion of memory is reserved for the VMM itself, and protected from any access by the virtual machine through segment truncation (see §3.2.3).

3.3   XEN—THE PARAVIRTUALIZATION ALTERNATIVE

Despite its many optimizations, VMware Workstation was viewed as an explicit trade-off between equivalence and performance, and indeed performed poorly on many micro benchmarks and some system-intensive workloads [27]. The use of dynamic binary translation and of shadow paging—

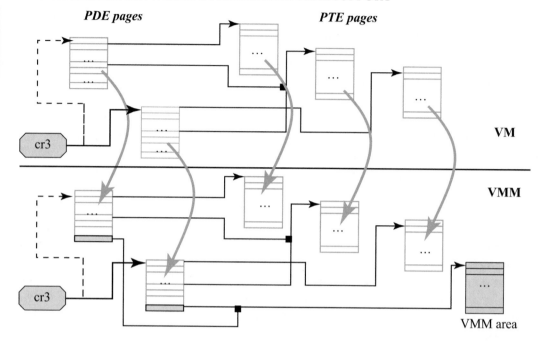

Figure 3.4: Using shadow page tables to virtualize memory [45]. The VMware VMM individually shadows pages of the virtual machine and constructs the actual linear address space used by the hardware. The top-most region is always reserved for the VMware VMM itself.

required because of its equivalence goal—introduced a degree of atypical complexity to the current system software.

Paravirtualization refers to hypervisors that sacrifice certain aspects of the equivalence property for a higher degree of efficiency or scalability. The term was introduced by the **Denali** isolation kernel in 2002 [182], a system designed to runs hundreds of concurrent **lightweight virtual machines**, each with a single address space. The term was the popularized by the **Xen** hypervisor [27]. Xen is the most influential open-source type-1 hypervisor for x86 architectures. Since its first release in 2003, Xen has gone through numerous iterations, in particular to take advantage of hardware support for virtualization.

In its original form on x86-32 architectures, Xen relied on paravirtualization to simply undefine all of the 17 non-virtualizable instructions of §2.5.2 [151], which must not be used. As a replacement, Xen defines an alternate interrupt architecture which consists of explicit calls from the guest operating system to the hypervisor (called **hypercalls**) and additional memory-mapped system registers. The Xen hypervisor relies on this modified x86-32 architecture to run virtual machines using direct execution only. The architecture also exposes a simplified MMU architecture that directly, but safely, exposes host-physical memory to the guest operating system.

The guest operating system combines virtual-to-guest-physical mappings with the guest-physical to host-physical information readable from the hypervisor to generate virtual-to-host-physical mappings, which are then passed down to the hypervisor for validation.

Table 3.2 lists the paravirtualization x86 interface of Xen [27], i.e., the list of changes to the underlying ISA exposed to virtual machines. There are changes in terms of memory virtualization, CPU virtualization and device I/O. As I/O is discussed later in the text (Chapter 6), we limit the discussion here to the first two topics.

Table 3.2: Xen's paravirtualized interface [27]

	Memory Management
Segmentation	Cannot install fully privileged segment descriptors and cannot overlap with the top end of the linear address space.
Paging	Guest OS has direct read access to hardware page tables, but updates are batched and validated by the hypervisor. A domain may be allocated discontinuous machine (aka host-physical) pages.
	CPU
Protection	Guest OS must run at a lower privilege level than Xen.
Exceptions	Guest OS must register a descriptor table for exception handlers with Xen. Aside from page faults, the handler remains the same.
System calls	Guest OS may install a "fast" handler for system calls, allowing direct calls from an application into its guest OS and avoiding indirection through Xen on every call.
Interrupts	Hardware interrupts are replaced with a lightweight event system.
Time	Each guest OS has a timer interface and is aware of both "real" and "virtual" time.

Memory management: Like VMware Workstation, Xen relies on **segment truncation** for protection. The Xen hypervisor is the only entity that runs at `%cpl=0` and can configure the segment descriptor tables. All segments available to virtual machines are truncated to exclude the top 64 MB of the linear address space. Xen runs out of that protected 64 MB region.

Xen has a paravirtualized approach to the virtualization of the MMU: Each VM consists of a contiguous guest-physical address space and a discontinuous set of host-physical pages. Both namespaces are visible to the guest operating systems. The guest OS can directly read the hardware page table structure accessed by the hardware CPU with PDE and PTE defined as host-physical addresses. To change a mapping, the guest OS determines the new host-physical mapping, to be validated by the hypervisor. Validations can be batched for performance. This approach is different from VMware Workstation, which relies on memory tracing to implement shadow paging, and hides the host-physical names from the VM.

CPU: Xen's simply replaces the 17 non-virtualizable instructions [151] with corresponding hypercalls which allow the guest to explicitly communicate with the hypervisor, e.g., to setup segment descriptors, disable interrupts, receive interrupt notification, and transition between userspace and the kernel. Non-virtualizable instructions are undefined and should not be used. With these changes, the augmented x86 architecture meets the Popek and Goldberg criteria, and Xen is architected as a trap-and-emulate hypervisor, unlike VMware.

Xen uses the x86-32 rings for protection: Xen runs at %cpl=0. The guest operating systems run at %cpl=1 and applications at %cpl=3. Page protection bits (pte.us) are used to protect the guest OS from the applications. Through the combination of segment truncation and page protection bits, a single address space can safely combine the Xen hypervisor itself, the guest operating system and the currently running application.

To eliminate unnecessary transitions via the hypervisor, the ISA extensions include the ability to perform fast system calls directly transitioning from the application to the guest OS. This insight was later picked up by the virtualization extension to reduce software overheads.

3.4   DESIGNS OPTIONS FOR TYPE-1 HYPERVISORS

§3.2.4 described the role of the host operating system in VMware Workstation, a type-2 hypervisor. We now discuss the approaches to type-1 hypervisor design used by VMware ESX Server [7, 177], Xen [27], and Microsoft Virtual Server.

Figure 3.5 shows the two main approaches to building a type-1 hypervisor. In both cases, we first note that the hypervisor is running, at all times, on all CPUs of the system. This is a major architectural difference with the hosted architecture of VMware Workstation (see §3.2.4). VMware Workstation has an asymmetrical architecture with a world switch, which implies that no piece of software is loaded at all times in memory. In contrast, type-1 hypervisors are always present in memory, able to directly handle all processor traps and I/O interrupts. No world switch is required.

The primary difference between flavors of type-1 hypervisors is whether the device drivers are embedded within the hypervisor or whether they run in a distinct, schedulable entity.

VMware ESX Server provides high performance disk and network I/O by implementing an efficient path from the I/O requests from the applications running in the virtual machines down to the physical devices [7]. The drivers are part of the hypervisor. The same CPU and the same thread of control are involved in all interactions necessary to initiate I/O, which reduces latency.

In contrast, the Xen and Microsoft Virtual Server do not include device drivers. Instead, they rely on a distinct, schedulable entity, called "dom0", which runs with elevated privileges, and in particular the ability to run physical device drivers. dom0 is scheduled by Xen and its main

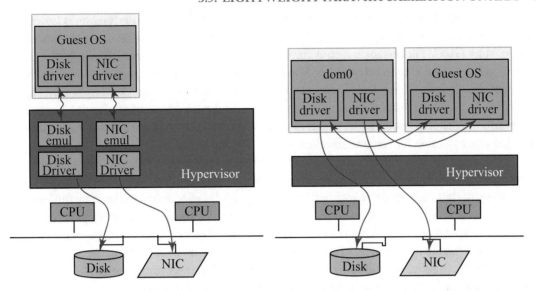

Figure 3.5: Two approaches to type-1 hypervisors: embedding drivers in the hypervisor (left) or scheduling them in dom0 (right).

function is to perform the back-end of I/O operations on behalf of all virtual machines. The virtual machines issue I/O request through asynchronous messages sent to dom0. This approach has the advantage of simplicity and portability as device drivers run in a standard operating system such as Linux (for Xen) or Windows (for Virtual Server). But it requires two distinct threads of control. In practice, the virtual machine and dom0 typically run on distinct cores of the system, and can take advantage of the natural overlap and asynchrony in I/O operations.

3.5   LIGHTWEIGHT PARAVIRTUALIZATION ON ARM

A downside to paravirtualization is that it can require detailed understanding of the guest operating system kernel to know how to modify its source code, and then requires ongoing maintenance and development to maintain potentially heavily modified versions of operating systems that can be run in virtual machines. Furthermore, the modifications are both architecture and operating system dependent, which only further add to the maintenance burden.

An alternative approach is **lightweight paravirtualization** [59]. Lightweight paravirtualization is a script-based method to automatically modify the source code of a guest operating system kernel to issue calls to the hypervisor instead of issuing sensitive instructions to enable a *trap-and-emulate* virtualization solution. Lightweight paravirtualization is architecture specific, but operating system independent. It is completely automated and requires no knowledge or understanding of the guest operating system kernel code.

Lightweight paravirtualization differs from Xen's paravirtualization as it requires no knowledge of how the guest is engineered and can be applied automatically on any OS source tree compiled by GCC. For instance, Xen defines a whole new file in `arch/arm/mm/pgtbl-xen.c`, which contains functions based on other Xen macros to issue hypercalls regarding memory management. Instead of existing kernel code, preprocessor conditionals are used to implement calls to these Xen functions in many places in the kernel code. Lightweight paravirtualization completely maintains the original kernel logic, which drastically reduces the engineering cost and makes the solution more suitable for test and development of existing kernel code.

However, there is a performance tradeoff between lightweight paravirtualization and traditional paravirtualization. The former simply replaces sensitive instructions with traps to the hypervisor, so that each sensitive instruction now traps and is effectively emulated by the hypervisor. If there are lots of sensitive instructions in critical paths, this can result in frequent traps and if traps are expensive, this can become a performance bottleneck. Traditional paravirtualization approaches typically optimize this further by replacing sections of code that may have multiple sensitive instructions with one paravirtualized hypercall to the hypervisor instead of repeated traps on each individual sensitive instruction, thereby improving performance.

While the designers of Xen took a traditional paravirtualization approach for x86, a lightweight paravirtualization approach was viewed more favorably in the context of supporting virtualization on ARM platforms. Two early efforts at building ARM hypervisors, KVM for ARM [59] and VMWare's Mobile Virtualization Platform (MVP) [28], were based on lightweight paravirtualization to address the problem that the original ARM architecture was not virtualizable. The designers of these systems hypothesized that the trap costs on ARM were lower than on x86, and that ARM's sensitive instructions occurred less frequently in critical code paths compared to x86 code, making a lightweight paravirtualization approach attractive.

Let us take a closer look at how lightweight paravirtualization was done in KVM for ARM, which was originally built for version 5 of the ARM architecture. As will be discussed in detail in Chapter 4, KVM is a type-2 hypervisor integrated into Linux that was originally designed for architectures that had hardware support for virtualization. Although ARM originally did not have such architectural support, the dominance of Linux on ARM made KVM an attractive starting point for building an ARM hypervisor. Linux provided a large amount of widely supported software functionality that runs on ARM hardware. KVM for ARM was an approach that simplified the development of a hypervisor by combining the benefits of lightweight paravirtualization with existing Linux infrastructure support for ARM.

To avoid the problems with sensitive non-privileged instructions, lightweight paravirtualization is used to slightly modify the guest kernel source code. There was no need to worry about user space software as user space applications will execute in the same CPU mode as if they were executing directly on a physical machine. Sensitive instructions are not generated by standard C-compilers and are therefore only present in assembler files and inline assembly.

KVM for ARM's lightweight paravirtualization is done using an automated scripting method to modify the guest kernel source code. The script is based on regular expressions and has been tested on a number of kernel versions with success. The script supports inline assembler syntax, assembler as part of preprocessor macros, and, assembler macros.

It works by replacing sensitive non-privileged instructions with trap instructions and emulating the sensitive instruction in software when handling the trap. However, KVM for ARM must be able to retrieve the original sensitive instruction including its operands to be able to emulate the sensitive instruction when handling a trap. KVM for ARM accomplishes this by defining an encoding of all the sensitive non-privileged instructions and their operands into trap instructions.

The SWI instruction on ARM always traps and is normally used for making system calls. The instruction contains a 24-bit immediate field (the payload), which can be used to encode sensitive instructions. Unfortunately, the 24 bits are not quite enough to encode all the possible sensitive non-privileged instructions and their operands. Some additional instructions that trap are needed for the encoding.

The ARM architecture defines **coprocessor access instructions** which are used to access the **coprocessor interface**. This interface does not relate to an actual physical coprocessor, but is merely used to extend the instruction set by transferring data between general purpose registers and registers belonging to one of the sixteen possible coprocessors. For example, the architecture always defines coprocessor number 15 which is called the *system control coprocessor* and controls the virtual memory system. Coprocessor access instructions are sensitive and privileged, so they will always trap, even for coprocessors that are not defined by the architecture. Only a few of the coprocessors are defined and in use, so those that are not defined can be repurposed for encoding sensitive non-privileged instructions. Specifically, coprocessors zero through seven are not in use and repurposed by KVM for ARM by using the coprocessor load/store access instruction. Taken together, the coprocessor load/store access instruction has 24 bits for its operands which can be leveraged to encode the sensitive non-privileged instructions. The combination of the SWI and coprocessor access instructions is sufficient to encode all possible sensitive non-privileged instructions and their operands.

The VMM needs to be able to distinguish between guest system calls and traps for sensitive instructions. KVM for ARM makes the assumption that the guest kernel does not make system calls to itself. Under this assumption, if the virtual CPU is in privileged mode, the payload is simply interpreted and the encoded instruction is emulated. If the virtual CPU is in user mode, the SWI instruction is considered as a system call made by the guest user space to the guest kernel.

As discussed in §2.5.3, the ARM architecture defines 24 sensitive non-privileged instructions in total. KVM for ARM encodes the instructions by grouping them in 15 groups; some groups contain many instructions and some only contain a single instruction. The upper 4 bits in the SWI payload indexes which group the encoded instruction belongs to (see Table 3.3). This leaves 20 bits to encode each type of instruction. Since there are 5 status register access functions and they need at most 17 bits to encode their operands, they can be indexed to the same type and be sub-indexed using additional 3 bits. There are 12 sensitive data processing instructions and they all use register 15 as the destination register and they all always have the S bit set (otherwise they are not sensitive). They are indexed in two groups: one where the I bit is set and one where it's clear. In this way, the data processing instructions need only 16 bits to encode their operands leaving 4 bits to sub-index the specific instruction out of the 12 possible. The sensitive load/store multiple and load/store with translation instructions are using 12 of the remaining 13 index values as can be seen in Table 3.3.

Table 3.3: Sensitive instruction encoding types

Index	Group/Instruction
0	Status register access instructions
1	LDM (2), P-bit clear
2	LDM (2), P-bit set
3	LDM (3), P-bit clear and W-bit clear
4	LDM (3), P-bit set and W-bit clear
5	LDM (3), P-bit clear and W-bit set
6	LDM (3), P-bit set and W-bit set
7	STM (2), P-bit set
8	STM (2), P-bit clear
9	LDRBT, I-bit clear
10	LDRT, I-bit clear
11	STRBT, I-bit clear
12	STRT, I-bit clear
13	
14	Data processing instructions, I-bit clear
15	Data processing instructions, I-bit set

In Table 3.3 only the versions of the load/store instructions with the I-bit clear are defined. This is due to a lack of available bits in the SWI payload. The versions with the I-bit set are encoded using the coprocessor load/store access instruction. When the I-bit is set, the load/store address is specified using an immediate value which requires more bits than when the I-bit is clear. Since the operands for coprocessor access instructions use 24 bits, 2 bits can be used to distinguish between the 4 sensitive load/store instructions. That leaves 22 bits to encode the instructions with the I-bit set, which is exactly what is needed.

An example may help illustrate the implementation of the KVM for ARM solution. Consider this code in `arch/arm/boot/compressed/head.S`:

```
mrs    r2, cpsr    @ get current mode
tst    r2, #3      @ not user?
bne    not_angel
```

The MRS instruction in line one is sensitive, since when executed as part of booting a guest, it will simply return the hardware CPSR. However, KVM for ARM must make sure that it returns the virtual CPSR instead. Thus, it can be replaced with a SWI instruction as follows:

```
swi    0x022000    @ get current mode
tst    r2, #3      @ not user?
bne    not_angel
```

When the SWI instruction in line one above generates a trap, KVM for ARM loads the instruction from memory, decodes it, emulates it, and finally returns to line two.

KVM for ARM using lightweight paravirtualization was implemented for ARMv5 and ARMv6, but never merged into the mainline Linux kernel. With the introduction of hardware virtualization support in ARMv7, a new KVM/ARM was designed to leverage ARM hardware virtualization extensions as discussed in Chapter 7, which became the fully supported Linux ARM hypervisor now available in mainline Linux.

3.6   FURTHER READING

The original VMware Workstation hypervisor is described in detail in [45]. Other papers describe key aspects of VMware Workstation, including its I/O and networking performance [162], the virtualization of its GPU [66]. Since their introduction, VMware Workstation has matured and successfully transitioned to take advantage of the emerging architectural support for virtualization of VT-x and AMD-v. Agesen et al. describe the evolution of VMware's approach to virtualization (see [4]).

Similarly, Pratt et al. describe the evolution of Xen, and in particular the transition from paravirtualization to full-virtualization (on machines with hardware support for it) [146]. Chisnal's 2007 textbook "The Definitive Guide to the Xen Hypervisor" provides a detailed description of the internals of Xen, both before and after the adoption of hardware virtualization [50].

VMware ESX Server (now called vSphere) is a commercial type-1 hypervisor. It shares the same VMware VMM subsystem as VMware Workstation, but packaged as part of a type-1 hypervisor that schedules I/O, CPU, and memory resources directly [7, 82, 177] and can live migrate VMs [135]. We particularly recommend Waldspurger's description of ESX memory management [177], which has been recognized by an ACM SIGOPS Hall of Fame Award [2].

KVM for ARM is described in more detail in [59] and VMware's MVP is described in [28]. An abandoned port of Xen for ARM [91] required comprehensive modifications to the guest kernel, and was never fully developed. None of these paravirtualization approaches could run unmodified guest OSes. These approaches have been superseded by solutions leveraging ARM hardware virtualization support, first introduced in ARMv7, as will be discussed in Chapter 7.

CHAPTER 4

x86-64: CPU Virtualization with VT-x

We now describe in a sequence of three chapters the architectural support for virtualization in x86-64 processors. This architectural support is the combination of innovation in the CPU (Chapter 4), MMU (Chapter 5), and I/O subsystem (Chapter 6).

This chapter describes VT-x, the Intel technology that virtualizes the CPU itself.[1] §4.1 first describes the key requirements set forth by Intel engineers in designing VT-x. §4.2 describes the approach to CPU virtualization, the concept of root and non-root modes, and how the architecture relates to the Popek and Goldberg theorem. §4.3 uses KVM—the Linux kernel virtual machine—as its case study on how to build a hypervisor that is explicitly designed to assume hardware support for virtualization. §4.4 discusses the performance implications of CPU virtualization, and in particular the implementation cost of atomic transitions between modes. Finally, like all chapters, we close with pointers for further reading.

4.1 DESIGN REQUIREMENTS

Intel Virtualization Technology, generally known as VT-x [171], was introduced in 2005 with the primary objective to provide architectural support for virtualization.

> A central design goal for Intel Virtualization Technology is to eliminate the need for CPU paravirtualization and binary translation techniques, and thereby enable the implementation of VMMs that can support a broad range of unmodified guest operating systems while maintaining high levels of performance.
>
> R. Uhlig et al., *IEEE Computer*, 2005 [171]

In stating this goal, Intel engineers observed that existing techniques such as paravirtualization and dynamic binary translation faced some hard technical challenges in virtualizing the existing x86 architecture. Uhlig et al. [171] lists the following specific challenges to approaches using software techniques.

---

[1]AMD's approach to virtualization, called AMD-v, is architecturally similar to VT-x.

- **Ring aliasing and compression:** guest kernel code designed to run at %cpl=0 must use the remaining three rings to guarantee isolation. This creates an alias as at least two distinct guest levels must use the same actual ring. Yet, software running in these two rings must be protected from each other according to the specification of the architecture.

- **Address space compression:** the hypervisor must be somewhere in the linear address space in a portion that the guest software cannot access or use.

- **Non-faulting access to privileged state:** some of the infamous **Pentium 17** [151] instructions provide read-only access to privileged state and are behavior-sensitive, e.g., the location in linear memory of the interrupt descriptor table (sidt), global descriptor table (sgdt), etc. These structures are controlled by the hypervisor and will be in a different location than the guest operating system specifies.

- **Unnecessary impact of guest transitions:** to address the efficiency criteria, it is essential that sensitive instructions (which must trigger a transition) be rare in practice. Unfortunately, modern operating systems rely extensively on instructions that are privilege-level sensitive, e.g., to suspend interrupts within critical regions, or to transition between kernel and user mode. Ideally, such instructions would not be sensitive to virtualization.

- **Interrupt virtualization:** the status of the interrupt flag (%eflags.if) is visible to unprivileged instructions (pushf). Since the flag can never be cleared when the guest is running as this would violate safety, any direct use of that instruction by the guest will lead to incorrect behavior. Furthermore, the popf instruction is control-sensitive as its semantic differs based on whether the CPU can control interrupts (i.e., whether %cpl$\leq$ %eflags.iopl).

- **Access to hidden state:** the x86-32 architecture includes some "hidden" state, originally loaded from memory, but inaccessible to software if the contents in memory changed. For example, the 32-bit segment registers are hidden: they are loaded from memory into the processor, but cannot be retrieved from the processor back into a general-purpose register or memory. As long as the contents in memory does not change, a segment register can be reloaded back into the processor, e.g., after a trap resulting from virtualization and hidden from the guest operating system. However, should the memory content ever change, that segment register can never be reloaded. Unfortunately, some legacy operating systems, and in particular Windows 95, modify the content of the segment descriptor table in some critical regions and yet explicitly rely on these very specific x86-32 semantics.

Of course, the evidence from the systems described in Chapter 3 suggests that these challenges can be mitigated in practice, but not entirely eliminated. For example, the address space compression challenge impacts performance whenever the top of the linear memory is used by the virtual machine: some guest operating systems may perform very poorly as a consequence. More concerning, a subset of the infamous "Pentium 17" instructions always return incorrect results to applications running at user-level, a limitation to equivalence [45].

In designing VT-x, Intel's central design goal is to fully meet the requirements of the Popek and Goldberg theorem, with the explicit goal that virtual machines running on top of a VT-x-based hypervisor meet the following three core attributes of equivalence, safety, and performance.

Equivalence: Intel's architects designed VT-x to provide absolute architectural compatibility between the virtualized hardware and the underlying hardware, which was furthermore to be backward-compatible with the legacy x86-32 and x86-64 ISA. This is much more ambitious than VMware Workstation, which was pragmatically focused on a set of well-defined guest operating systems, and of paravirtualization approaches such as Xen, which required kernel modifications and in practice only applied to open-source operating systems.

Safety: Prior hypervisors based on dynamic binary translation or paravirtualization provided security and isolation, but through a reasoning that involved complex invariants that had to be maintained in software, such as the correct use of segmentation for protection. Through architectural support specifically designed for virtualization, a much-simplified hypervisor can provide the same characteristics with a much smaller code base. This reduces the potential attack surface on the hypervisor and the risk of software vulnerabilities.

Performance: Ironically, an increase in performance over existing, state-of-the-art virtualization techniques was **not** a release goal with the first-generation hardware support for virtualization. Instead, the goal at the beginning was merely to setup the appropriate architecture and a roadmap for ongoing improvements at the architectural and micro-architectural level. Indeed, the first-generation processors with hardware support for virtualization were not competitive with state-of-the-art solutions using DBT [3].

4.2   THE VT-X ARCHITECTURE

The architecture of VT-x is based on a central design decision: do **not** change the semantics of individual instructions of the ISA. This includes of course the instructions that most obviously violate the virtualization principle. As a corollary, the architecture does **not** attempt to separately address the individual aspects of the architecture that limited virtualization, as described in §4.1. Instead, VT-x **duplicates** the entire architecturally visible state of the processor and introduces a new mode of execution: the **root mode**. Hypervisors and host operating systems run in **root mode** whereas virtual machines execute in **non-root mode**. This architectural extension has the following properties.

- The processor is at any point in time either in root mode or in non-root mode. The transitions are atomic, meaning that a single instruction or trap can transition from one mode to the other. This differs notably from the conventional implementation of a context switch by an operating system, which requires a convoluted instruction sequence.

- The root mode can only be detected by executing specific new instructions, which are only available in root mode. In particular, it cannot be inferred through any other mechanisms or memory access. This is necessary to ensure that root-mode execution itself can be virtualized and to support recursive virtualization.

- This new mode (root vs. non-root) is only used for virtualization. It is orthogonal to all other modes of execution of the CPU (e.g., real mode, v8086 mode, protected mode), which are available in both modes. It is also orthogonal to the protection levels of protected mode (e.g.%cpl=0--%cpl=3) with all four levels separately available to each mode.

- Each mode defines its own distinct, complete 64-bit linear address space. Each address space is defined by a distinct page table tree with a distinct page table register. Only the address space corresponding to the current mode is active in the TLB, and the TLB changes atomically as part of the transitions.

- Each mode has its own interrupt flag. In particular, software in non-root mode can freely manipulate the interrupt flags (%eflags.if). External interrupts are generally delivered in root mode and trigger a transition from non-root mode if necessary. The transition occurs even when non-root interrupts are disabled.

Figure 4.1 illustrates how the central design of VT-x is intended to be used by system software: in root mode, software has access to the full (non-virtualized) architecture of x86-64, including all privilege rings of protected mode (shown in the figure) as well as the legacy modes of execution (not shown). This provides backward compatibility in the architecture, and consequently for software, e.g., a host operating system will typically run in root-%cpl=0 and its application programs will run in root-%cpl=3. The hypervisor also runs in root-%cpl=0 where it can issue new privileged instructions to enter into non-root mode. Figure 4.1 also shows that virtual machines execute with the full duplicate of privilege rings: each guest operating system runs in non-root-%cpl=0 and applications run in non-root-%cpl=3.

4.2.1   VT-X AND THE POPEK/GOLDBERG THEOREM

Recall Popek and Goldberg's central virtualization theorem, discussed in §2.2.

> Theorem 1 [143]: For any conventional third-generation computer, a virtual machine monitor may be constructed if the set of sensitive instructions for that computer is a subset of the set of privileged instructions.

The VT-x architecture meets the criteria of the theorem, but through a significant departure from the original model proposed to demonstrate the theorem. Popek and Goldberg identified the key criteria for virtualization. Through their proof-by-construction of the theorem, they furthermore

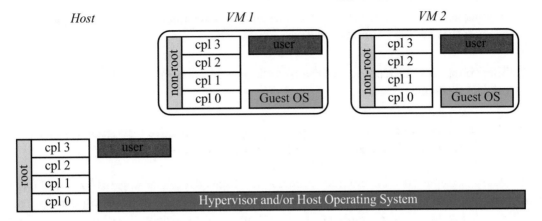

Figure 4.1: Standard use by hypervisors of VT-x root and non-root modes.

demonstrated that virtualization was achievable without requiring any additional hardware constraints than those necessary to support protected operating systems, and specifically by using address space compression and ring aliasing (running both the guest operating system and applications at user level) to build the hypervisor.

Intel pragmatically took a different path by duplicating the entire state of the processor and introducing the dedicated `root-mode`. In particular, the full duplication of the architecture was motivated by the need to ensure backward compatibility for the ISA and full equivalence for virtualized workloads [171]. Terms must therefore be redefined to convincingly express that VT-x follows the Popek/Goldberg criteria. The corresponding core VT-x design principle can be informally framed as follows.

> In an architecture with root and non-root modes of execution and a full duplicate of processor state, a hypervisor may be constructed if all sensitive instructions (according to the non-virtualizable legacy architecture) are root-mode privileged.
>
> When executing in non-root mode, all root-mode-privileged instructions are either (i) implemented by the processor, with the requirement that they operate exclusively on the non-root duplicate of the processor or (ii) cause a trap.

We make three observations.

1. Unlike the original Popek and Goldberg theorem, this rephrasing does not take into account whether instructions are privileged or not (as defined by their availability to software

running at %cpl>0), but instead only takes into consideration the orthogonal question of whether they are **root-mode privileged**.

2. These traps are sufficient to meet the equivalence and safety criteria. This is similar to the original theorem.

3. However, reducing transitions by implementing certain sensitive instructions in hardware is necessary to meet the performance criteria for virtualization.

A few examples help illustrate how the notion of sensitivity is orthogonal to that of (regular) privilege. First, a guest operating system running in non-root-%cpl=0 will issue certain privileged instructions such as reading or even writing control registers: these instructions must operate on the non-root duplicate of the processor state. Following the principle, two implementations are possible in this situation: (i) either execute the instruction in hardware onto the non-root context since the entire processor state is duplicated. The hypervisor must not be informed since the hardware fully handles the instructions; and (ii) or take a trap from non-root mode, which allows a hypervisor to emulate the instruction.

The former is preferred from a performance perspective as it reduces the number of transitions. Obviously, it requires specific architectural support by the processor. The implementation decision is therefore a tradeoff between hardware complexity and overall performance.

As a second example, the instructions that manipulate the interrupt flag (cli, sti, popf) are sensitive, in particular since other instructions can actually leak information about the interrupt flag itself onto memory (pushf). However, recall that these four instructions are **not** privileged instructions in the architecture, i.e., they are available to user-level applications, at least in some cases. Given their high execution rate in modern kernels, their non-root implementation is handled directly by the processor.

As a third example, some instructions available at user-level such as sgdt and sidt, are known to be behavior-sensitive. These rare instructions are not privileged, yet virtualization-sensitive. Therefore, they must be (and are) root-mode-privileged.

4.2.2   TRANSITIONS BETWEEN ROOT AND NON-ROOT MODES

Figure 4.2 shows the key interactions and transitions between root mode and non-root mode in the implementation of VT-x. The state of the virtual machine is stored in a dedicated structure in physical memory called the **Virtual Machine Control Structure (VMCS)**. Once initialized, a virtual machine resumes execution through a vmresume instruction. This privileged instruction loads the state from the VMCS in memory into the register file and performs an atomic transition between the host environment and the guest environment. The virtual machine then executes in non-root mode until the first trap that must be handled by the hypervisor or the next external interrupt. This transition from non-root mode to root mode is called a #vmexit.

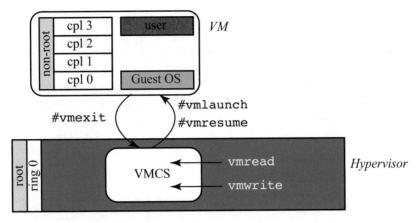

Figure 4.2: VT-x transitions and control structures.

Table 4.1 lists the various possible reasons for a #vmexit. The reason itself is stored in a dedicated register (vmcs.exit_reason) to accelerate emulation. The reasons for an exit can be grouped into categories including the following.

- Any attempt by the guest to execute a *root-mode-privileged* instruction, as this is the fundamental requirement of the Popek and Goldberg theorem. This includes most of the privileged instructions of the classic (i.e., non-virtualizable) x86 architecture, as well as the sensitive-yet-unprivileged instructions causing violations of the theorem.

- The new vmcall instructions, designed to allow explicit transitions between non-root mode and root-mode, and in particular hypercalls made by the guest operating system and destined to the hypervisor. This is analogous to the sysenter instruction that transitions between user mode and kernel mode.

- *Exceptions* result from the execution of any innocuous instruction in non-root mode, that happens to take a trap. This includes, in particular, page faults (#PF) caused by shadow paging, the access to memory-mapped I/O devices, or general-purpose faults (#GP) due to segment violations.

- *EPT violations* are the subset of page faults caused when the extended page mapping (under the control of the hypervisor) is invalid. This exit was introduced with extended paging (see Chapter 5).

- *External interrupts* that occurred while the CPU was executing in non-root mode, e.g., as the result of network or disk I/O on the host. Such events must be handled by the hypervisor (in the case of type-1 designs) or the host operating system (for type-2 designs) and may or may not have any side-effects to the virtual machine itself.

- The *interrupt window* opens up whenever the virtual machine has enabled interrupt and the virtual machine has a pending interrupt. Following the #vmexit, the hypervisor can emulate the pending interrupt onto the virtual machine.

- Finally, the ISA extensions introduced with VT-x to support virtualization are also control-sensitive and therefore cause a #vmexit, each with a distinct exit reason. Such exits never occur during "normal virtualization," but play a fundamental role in nested virtualization [33, 191].

Table 4.1: Categories VT-x exit codes

Category	Exit Reason	Description
Exception	0	Any guest instruction that causes an exception
Interrupt	1	The exit is due to an external I/O interrupt
Triple fault	2	Reset condition (bad)
Interrupt window	7	The guest can now handle a pending guest interrupt
Legacy emulation	9	Instruction is not implemented in non-root mode; software expected to provide backward compatibility, e.g., task switch
Root-mode Sensitive	11-17, 28-29, 31-32, 46-47:	x86 privileged or sensitive instructions: getsec, hlt, invd, invlpg, rdpmc, rdtsc, rsm, mov-cr, mov-dr, rdmsr, wrmsr, monitor, pause, lgdt, lidt, sgdt, sidt, lldt, ltr, sldt
Hypercall	18	vmcall : Explicit transition from non-root to root mode
VT-x new	19-27, 50, 53	ISA extensions to control non-root execution: invept, invvpid, vmclear, vmlaunch, vmptrld, vmptrst, vmreas, vmresume, vmwrite, vmxoff, vmxon
I/O	30	Legacy I/O instructions
EPT	48-49	EPT violations and misconfigurations

The transitions between root and non-root mode are architecturally atomic: a single instruction,— vmresume—transitions back to non-root mode and loads the VMCS state into the current processor state. In the other direction, the trap #vmexit stores the state of the virtual CPU into the VMCS state. Although the VMCS state is backed by a specific memory region, the architecture does not specify whether the processor must spill the entire state into the cached memory, or whether it can hold off to a subset into the processor itself. As a result, the in-memory state of the current VMCS is undetermined. In addition, the layout of the VMCS is undefined. The hypervisor software must access selected portions of the guest state via the vmread and vmwrite instruction pair.[2]

4.2.3 A CAUTIONARY TALE—VIRTUALIZING THE CPU AND IGNORING THE MMU

Today, all processor architectures with virtualization support also virtualize the MMU. This, however, has not always been the case. Indeed, the first-generation Intel CPU with VT-x provided only elementary support for memory virtualization. In such a design, root-mode and non-root mode each have a distinct %cr3 register specifying the base of the page table tree. The cr3 register is atomically updated as part of the vmentry and #vmexit transition between root and non-root mode. As a result, a hypervisor can configure 100% disjoint address spaces. This solves the address compression challenge of previous hypervisor and removes the need to rely on segmentation for protection of the hypervisor. But in such a minimal design, every other aspect of memory virtualization is left to software. As in prior architectures without any architectural support, two approaches remain possible: (1) to shadow the page tables of the virtual machine (with guest-physical values) with a duplicate set of page tables which contains the host-physical values as was done by VMware (see §3.2.5); or (2) to rely on paravirtualization of the virtual memory subsystem and inform the hypervisor to validate all new mappings (see §3.3).

This choice had a severe and unanticipated consequence when shadowing page tables. Using shadow paging, the first two generations of Intel processors failed to address the performance criteria: over 90% of #vmexit transitions were due to shadow paging and the resulting virtual machine performance was slower than when simply disabling VT-x and using software techniques [3].

The explanation for this anomaly is actually subtle: shadowing operates by keeping the mappings of the hardware page tables in sync with the changes made by the guest operating system in its own page tables. The mechanism to identify changes is the MMU itself: all guest page table pages are downgraded to read-only mappings to ensure a transition from the virtual machine to the hypervisor, i.e., a trap in classic architectures and a #vmexit in VT-x. A hypervisor that relies entirely on direct execution will therefore necessarily suffer a trap for every change in the virtual memory of the guest operating system. In contrast, VMware's memory tracing mechanism relied on **adaptive binary translation** to eliminate the overwhelming majority of page faults. Adaptive dynamic binary translation is an effective technique as it relies on the great locality of the instructions that manipulate page table entries, a handful per operating system. Once these instruction locations have been dynamically identified, the adaptive retranslation process simply

---

[2]In designing its own virtualization extension, AMD took a slightly different approach that architecturally defines the in-memory data structure which it calls the *virtual machine control block (VMCB)*. Even though ISAs differ, Intel's VMCS and AMD's VMCB each offer an equivalent, yet incompatible, abstraction of the virtual CPU state.

emulates the instructions and updates the shadow entries without ever dereferencing the memory location directly and therefore without taking an expensive trap.

Fortunately, this anomaly was addressed in subsequent processors.

4.3   KVM—A HYPERVISOR FOR VT-X

So far, we have described the hardware extensions introduced by VT-x, and discussed a few architectural considerations. We now use **KVM** [113], the Linux-based Kernel Virtual Machine, as a case study to put the innovation in practice. KVM makes for an interesting study because of its maturity and simplicity.

- KVM is the most relevant open-source type-2 hypervisor. It is used in numerous projects, cloud hosting solutions, and deployed in enterprises and private clouds. KVM is the officially supported hypervisor of major commercial Linux distributions and the foundation of most Openstack deployments.

- KVM relies on **QEMU** [32], a distinct open-source project, to emulate I/O. Absent KVM, QEMU is a complete machine simulator with support for cross-architectural binary translation of the CPU, and a complete set of I/O device models. Together with KVM, the combination is a type-2 hypervisor, with QEMU responsible for the userspace implementation of all I/O front-end device emulation, the Linux host responsible for the I/O backend (via normal system calls) and the KVM kernel module responsible to multiplex the CPU and MMU of the processor.

- The kernel component of KVM, which implements the functionality of CPU and memory virtualization equivalent to the VMware VMM, has been a built-in component of the Linux kernel and is designed to avoid unnecessary redundancies with Linux. KVM was designed from the beginning to be part of Linux, with the explicit goal to merge all kernel-resident components with the mainline Linux source tree. This was achieved with the merge of the KVM kernel module into the Linux mainline as of version 2.6.20 in 2007.

- Unlike Xen or VMware Workstation, KVM was designed from the ground up assuming the existence of hardware support for virtualization.[3] This makes for a particularly good study of the intrinsic complexity of a hypervisor designed for VT-x.

4.3.1   CHALLENGES IN LEVERAGING VT-X

The developers of KVM adapted Popek and Goldberg's three core attributes of a virtual machine as follows.

---

[3]Originally, KVM was designed for x86-64 processors with either VT-x or AMD-v; since then, it has evolved to virtualize other architectures notably ARM/ARM64. See §7.6.

Equivalence: A KVM virtual machine should be able to run any x86 operating system (32-bit or 64-bit) and all of its applications without any modifications. KVM must provide sufficient compatibility at the hardware level such that users can choose their guest operating system kernel and distribution.

Safety: KVM virtualizes all resources visible to the virtual machine, including CPU, physical memory, I/O busses and devices, and BIOS firmware. The KVM hypervisor remains in complete control of the virtual machines at all times, even in the presence of a malicious or faulty guest operating system.

Performance: KVM should be sufficiently fast to run production workloads. However, KVM's explicit design of a type-2 architecture implies that resource management and scheduling decisions were left as part of the host Linux kernel (of which the KVM kernel module is a component). To achieve these goals, KVM's design makes a careful tradeoff to ensure that all performance-critical components are handled within the KVM kernel module while limiting the complexity of that kernel module. Specifically, the KVM kernel module handles only the core platform functions associated with the emulation of the x86 processor, the MMU, the interrupt subsystem (inclusive of the APIC, IOAPIC, etc.); all functions responsible for I/O emulation are handled in userspace.

To further simplify the implementation, the original version of KVM leveraged highly two existing open-source project: (i) QEMU [32] for all I/O emulation in user-space and (ii) the x86-specific portions of Xen, which served as the starting point for the Linux kernel module. Since then, KVM and Xen have followed largely disjoint paths, but the evolution of QEMU is mostly driven by the requirements for KVM.

4.3.2   THE KVM KERNEL MODULE

The kernel module only handles the basic CPU and platform emulation issues. This includes the CPU emulation, memory management and MMU virtualization, interrupt virtualization, and some chipset emulation (APIC, IOAPIC, etc.). But it excludes all I/O device emulation.

Given that KVM was designed only for processors that follow the Popek/Goldberg principles, the design is in theory straightforward: (i) configure the hardware appropriately; (ii) let the virtual machine execute directly on the hardware; and (iii) upon the first trap or interrupt, the hypervisor then regains control, and "*just*" emulates the trapping instruction according to the semantic.

The reality is much more complex. Our analysis is based on the linux-4.8 release of October 2016. The KVM kernel module alone has over 25,000 source lines of code (LOC). The complexity is due in part to: (i) the need to support multiple distinct versions of VT-x going back to the first processor with VT-x (Intel's *Prescott*); (ii) the inherent complexity of the x86 instruction set architecture; and (iii) the remaining lack of architectural support in hardware for some basic operations, which must be handled in software.

Figure 4.3 illustrates the key steps involved in the trap handling logic of KVM, from the original #vmexit until the vmresume instruction returns to non-root mode. Immediately

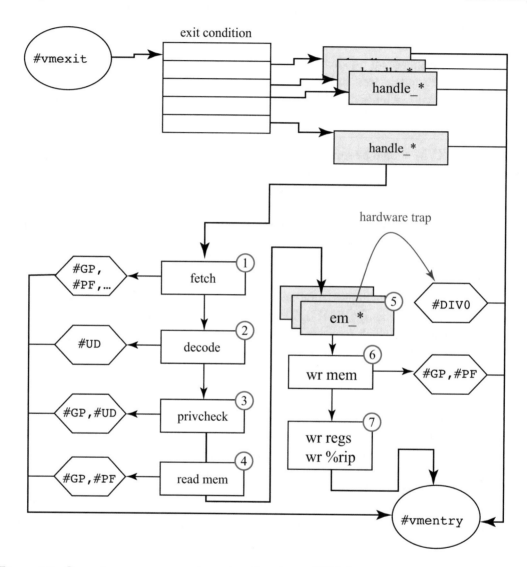

Figure 4.3: General-purpose trap-and-emulate flowchart of KVM.

upon the #vmexit, KVM first saves all the vcpu state in memory. The Intel architectural manual [103](Appendix I) defines 54 possible exit reasons (see Table 4.1). KVM then performs a first-level dispatch based on vmcs.exit_reason, with a distinct handler (handler_* in arch/x86/kvm/vmx.c) for each exit reason. Most of these handlers are straightforward. In particular, some common code paths rely exclusively on VMCS fields to determine the necessary emulation steps to perform. Depending on the situation, KVM may:

- emulate the semantics of that instruction and increment the instruction pointer to the start of the next instruction;

- determine that a fault or interrupt must be forwarded to the guest environment. Following the semantic of x86, KVM changes the stack pointer and stores on the stack the previous instruction and stack pointers. Execution will then resume at the instruction specified by the guest's interrupt descriptor table;

- change the underlying environment and re-try the execution. This occurs for example in the case of an EPT violation; and

- do nothing (at least to the virtual machine state). This is the case for example when an external interrupt occurs, which is handled by the underlying host operating system. Non-root execution will eventually resume where it left off.

Unfortunately, the information available in the VMCS is at times inadequate to handle the #vmexit without actually decoding the instruction that caused it. KVM—as well as any hypervisor for VT-x which attempts to provide equivalence—must therefore also include a general-purpose decoder, capable of decoding all instructions, and a general-purpose emulator, which can emulate them. Figure 4.3 illustrates the key steps involved in the general-purpose emulator implementation of KVM. The core logic that performs this basic step is implemented in arch/x86/kvm/emulate.c, a file with 5000+ lines of code filled with macros, complexity, and subtleties. The key steps of the emulator are:

1. **fetch** the instruction from guest virtual memory (%cs:%eip). First, the virtual address must be converted into a linear address, then into a guest-physical address, and finally fetched from memory;

2. **decode** the instruction, extracting its operator and operands. The CISC nature and variable length of the x86-64 instructions makes this process non-trivial;

3. **verify** whether the instruction can execute given the current state of the virtual CPU, e.g., privileged instructions can only execute if the virtual CPU is at cpl0;

4. **read** any memory read-operands from memory, as is the common case in the x86 CISC architecture using the same virtual to linear to guest-physical relocation steps as for instruction fetches;

5. **emulate** the decoded instruction, which could be any of the instructions defined in the x86 architecture. Each instruction opcode is emulated via its own dedicated emulation routine (em_* in the source);

6. **write** any memory write-operands back to the guest virtual machine; and

7. **update** guest registers and the instruction pointer as needed.

Clearly, these steps are complex, expensive, and full of corner cases and possible exception conditions. Figure 4.3 further shows the intricacy as every stage contains a possible exception case, in which the trap-and-emulate logic concludes that the guest instruction *cannot* successfully execute, and instead should generate a fault in the guest virtual CPU, e.g., a virtual #GP, #PF, or #UD. Furthermore, the instruction emulation step can, in some rare cases, lead to a fault in the actual hardware such as when dividing by zero. This is shown in red in the figure, leading to additional complexity.

The emulator is also notoriously brittle, and its implementation has evolved over the years to address defect reports as well as ongoing ISA extensions. It remains error-prone. A recent study by Amit et al. [15] identified 117 emulation bugs in the KVM kernel module, of which 72 were in the emulator alone.

4.3.3 THE ROLE OF THE HOST OPERATING SYSTEM

KVM was specifically designed to be part of Linux. Unlike other type-2 hypervisors such as VMware Workstation [45] or VirtualBox [180] which are host-independent, KVM is deeply integrated into the Linux environment. For example, the `perf` toolkit has a specific mode to profile KVM virtual machines; this is made possible by the deep integration of the KVM and Linux projects.

Figure 4.4 shows the core KVM virtual machine execution loop [113], shown for one virtual CPU. The outer loop is in usermode and repeatedly:

- enters the KVM kernel module via an `ioctl` to the character device `/dev/kvm`;

- the KVM kernel module then executes the guest code until either (i) the guest initiates I/O using an I/O instruction or memory-mapped I/O or (ii) the host receives an external I/O or timer interrupt;

- the QEMU device emulator then emulates the initiated I/O (if required); and

- in the case of external I/O or timer interrupt, the outer loop may simply return back to the KVM kernel module by using another `ioctl`(`/dev/kvm`) without further side-effects. This step in userspace is however essential as it provides the host operating system with the opportunity to make global scheduling decisions.

The inner loop (within the KVM kernel module) repeatedly:

- restores the current state of the virtual CPU;

- enters `non-root` mode using the `vmresume` instruction. At that point, the virtual machine executes in that mode until the next `#vmexit`;

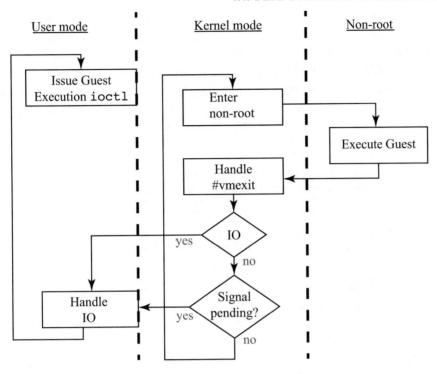

Figure 4.4: KVM Virtual Machine Execution Loop, adapted from [113].

- handles the #vmexit according to the exit reason, as described in §4.3.2;

- if the guest issued a programmed IO operation (exit_reason = IO) or a memory-mapped IO instruction (exit_reason = exception, but only when accessing a memory-mapped IO page), break the loop and return to userspace; and

- if the #vmexit was caused by an external event (e.g., exit_reason = interrupt), break the loop and return to userspace.

4.4   PERFORMANCE CONSIDERATIONS

The design of VT-x is centered around the duplication of architectural state between root and non-root modes, and the ability to atomically transition between them: a single instruction, vmresume, transitions back to non-root mode and loads then VMCS state into the current processor state. In the other direction, the trap #vmexit stores the entire state of the virtual CPU into the VMCS state.

Atomic transitions between modes do not imply a high execution speed, and certainly not a single-cycle execution time. Intuitively, such transitions are expected to stall the entire execution pipeline, given that the whole register file, privileged state, and instruction pointer change. In reality, the measured cost of these transitions is high, and suggests a complex implementation within the processor's microcode firmware. Table 4.2 shows the cost of a hardware round-trip, defined by a #vmexit followed by a NULL handler in the hypervisor that increments the instruction pointer and resumes execution of the virtual machine. In early generations, such as Prescott, the cost of a single hardware round-trip was measured in microseconds, largely exceeding the cost of a regular trap. Since then, transitions have improved by 5× but remain high.

Table 4.2: Hardware round-trip latency, from [5]

Microarchitecture	Launch Date	Cycles
Prescott	3Q05	3963
Merom	2Q06	1579
Penryn	1Q08	1266
Nehalem	3Q09	1009
Westmere	1Q10	761
Sandy Bridge	1Q11	784

Table 4.3 further shows the cost of individual VT-x instructions and transitions for a broader set of processors, including more recent releases [6]. Despite the lack of description by Intel or an independent study of the cause of the changes, one can reasonably infer that certain instructions, (e.g., vmread) went from being implemented in microcode to being integrated directly into the pipeline. Nevertheless, all the atomic transitions (vmresume or #vmexit) remain very expensive despite having been highly optimized, and have not improved noticeably in the past five generations of processors.

4.5   FURTHER READING

Amit et al. [15] identified minor hardware limitations to the claim that VT-x is virtualizable according to the Popek/Goldberg theorem: the physical address width is available directly through the CPUID instruction and not virtualizable. This can cause a problem when live-migrating a virtual machine [51, 135] across systems with different physical address widths. Also, some of the FPU state is not fully virtualizable, and workarounds are required.

In the same work, Amit et al. primarily identified software bugs in the implementation of KVM that violate the safety equivalence property expected of hypervisors. By comparing systematically the behavior of a KVM virtual machine with Intel's reference simulator [15], they identified 117 distinct bugs in KVM. The common cause for these bugs is that the entire complexity of the x86 ISA is exposed to software. Although the majority of these bugs are corner-case equivalence limitations with little practical impact, at least 6 bugs have led to security vulnerabilities and can cause host DoS, guest DoS, or privilege escalation. These concerns have called for

Table 4.3: Hardware costs of individual VT-x instructions and #vmexit for different Intel processors

Processor	Prescott	Merom	Penryn	Westmere	Sandy Bridge	Ivy Bridge	Haswell	Broadwell
VMXON	243	162	146	302	108	98	108	116
VMXOFF	175	99	89	54	84	76	73	81
VMCLEAR	277	70	63	93	56	50	101	107
VMPTRLD	255	66	62	91	62	57	99	109
VMPTRST	61	22	9	17	5	4	43	44
VMREAD	178	53	26	6	5	4	5	5
VMWRITE	171	43	26	5	4	3	4	4
VMLAUNCH	2478	948	688	678	619	573	486	528
VMRESUME	2333	944	643	402	460	452	318	348
vmexit/vmcall	1630	727	638	344	365	334	253	265
vmexit/cpuid	1599	764	611	389	434	398	327	332
vmexit/#PF	1926	1156	858	569	507	466	512	531
vmexit/IOb	1942	858	708	427	472	436	383	397
vmexit/EPT	N/A	N/A	N/A	546	588		604	656

a refactoring of the KVM kernel module, and in particular to consider moving the instruction emulator to userspace where bugs can be more easily contained [37].

Readers interested in getting a deeper understanding of KVM will quickly realize that the source code is the best form of documentation, even though the KVM website [118] does contain some useful pointers.

All hypervisors for the x86 architecture have embraced VT-x since its introduction in silicon. Current versions of VMware Workstation leverages the VT-x MMU capabilities similarly to KVM: the world switch no longer exists in its original form, and the core execution loop of VMware Workstation resembles that of KVM. Adams and Agesen [3] study in great detail this tradeoff between DBT and direct execution in the *presence* of hardware support. The tradeoff is substantial whenever the hypervisor must shadow in-memory data structures that are frequently accessed in practice. The tradeoff was fundamental prior to the introduction of extended paging in processors, and the authors conclude that VT-x (without extended paging) hurts performance.

Xen also embraced VT-x early under the term **hardware virtualization (HVM)** [50, 146]. With HVM, guest operating systems no longer need to be paravirtualized to run on Xen. Instead, paravirtualization is merely an optional set of extensions that improve performance and functionality by directly communicating with the hypervisor.

To the best of our knowledge, the micro-architectural cost of transitions from root to non-root mode has never been academically studied in depth, and the salient aspects of their implementation in Intel or AMD processors have not been disclosed. The cause of the 5× improvement from Prescott to Sandy Bridge remains the cause for speculation. Instead, the focus has been to relentlessly reduce the software cost of handling a `#vmexit`.

CHAPTER 5

x86-64: MMU Virtualization with Extended Page Tables

Hypervisors must virtualize physical memory, so that each virtual machine has the illusion of managing its own contiguous region of physical memory. Recall the definitions of §1.6: each virtual machine is provided the abstraction of **guest-physical memory**, while the hypervisor manages **host-physical memory**, the actual underlying physical resource.

This creates a two-dimensional problem: the guest operating system defines mapping between virtual memory and guest-physical memory. The hypervisor then independently defines mappings between guest-physical memory and host-physical memory.

In the absence of any architecture support in the memory management unit, hypervisors rely on **shadow paging** to virtualize memory. In shadow paging, the hypervisor manages a composite set of page tables that map virtual memory to host-physical memory (see §3.2.5 for a description of VMware's implementation). Shadow paging, as implemented in software, is arguably the most complex subsystem of a hypervisor. It relies on memory tracing to keep track of the changes to page table structures in memory. Shadow paging also relies heavily on heuristics to determine which pages should be traced, as page tables can be anywhere in physical memory and can be allocated and reallocated at the discretion of the guest operating system. Further, the introduction of VT-x had a negative impact on the performance of shadow paging (see §4.2.3).

Extended page tables provide architectural support for MMU virtualization. §5.1 describes the design of extended paging in x86-64 processors. §5.2 describes how KVM manages and virtualizes memory, and takes advantage of extended page tables. §5.3 measures the cost of MMU virtualization. Finally, like all chapters, we close with pointers for further reading.

5.1 EXTENDED PAGING

Extended page tables, also known as **nested page tables**, eliminates the need for software-based shadow paging. The design was published by Bhargava et al. [35] in 2008 and available in silicon by both AMD and Intel around that same time.

Extended page tables combine in hardware the classic hardware-defined page table structure of the x86, maintained by the guest operating system, with a second page table structure, maintained by the hypervisor, which specifies guest-physical to host-physical mappings. Both structures are similarly organized as a tree.

With extended paging, the TLB lookup logic, which is deeply integrated with the processor, does not change: the TLB remains organized as a set associative cache that map virtual pages to host-physical pages. On architectures that support superpages (e.g., 2 MB and 1 GB on x86-64), there is typically a distinct cache for each page size.

What changes fundamentally is the TLB miss handling logic. Whenever a mapping is not present in the TLB, the x86-64 architecture specifies that the processor will walk the page table tree, rooted by %cr3 and insert the missing mapping in the TLB. On x86-64, in the absence of extended paging, the tree is a $n = 4$ level tree for regular pages, $n = 3$ for 2 MB pages, and $n = 2$ for 1 GB pages. The page walk logic therefore must access n locations in memory to find the missing mapping. Only then can the CPU perform the actual memory read or write.

Let's consider the case where the virtual to guest-physical tree is n-level and the guest-physical to host-physical tree is m-level. This is shown in Figure 5.1. On TLB misses, the hardware walks the guest page table structure which consists entirely of guest-physical pages, with each guest-physical reference required to be individually mapped to its own host-physical address. To resolve each reference in the first tree, the processor must first perform m references in the second tree, and then lookup the mapping in the first tree. There are n such steps, each requiring $m + 1$ references. These $n \times (m + 1)$ references lead to the desired guest-physical address. Another m lookups are required to convert this guest-physical address into a host-physical address. The number of architecturally defined memory references required for an extended page lookup is therefore $n \times m + n + m$.

In summary, extended paging composes the two independent mappings via a quadratic lookup algorithm. Although architecturally quadratic, current-generation processors rely extensively on secondary data structures to reduce the number of actual references to the memory hierarchy.

5.2   VIRTUALIZING MEMORY IN KVM

KVM was originally released prior to the introduction of extended paging. As a result, the KVM kernel module may be configured to enable or disable the feature [42]. In practice, however, any considerations to the use of KVM today (in 2017) without extended paging are limited to cases of nested virtualization, e.g., the Turtles research prototype [33].

Figure 5.2 shows the key aspects of memory management in KVM. The figure shows that there are **three** distinct page table structures, each managed by a distinct entity (visualized by the red dot).

- QEMU, as a userspace process, allocates the guest-physical memory in a contiguous portion of its own virtual address space. This is convenient for a number of reasons: (i) it lets the details of memory management and allocation to the host operating system, in line with a type-2 architecture; and (ii) it provides convenient access from userspace to the guest-physical address space; this is used in particular by emulated devices that perform DMA to and from the guest-physical memory. Like any process on the machine, Linux manages a

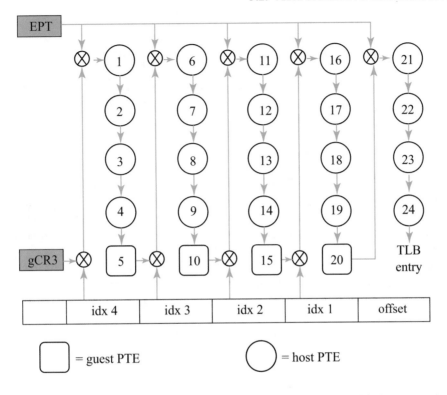

Figure 5.1: Sequence of architecturally-defined memory references to load a TLB entry with extended page tables ($m = 4$ and $n = 4$).

page table tree for this process. When the QEMU process is scheduled—which includes all situations where the virtual machine's VCPU is running—, the `root mode %cr3` register defines an address space that includes QEMU and the contiguous guest-physical memory.

- The virtual machine manages its own page tables. With nested paging enabled, the `non-root cr3` register points to the guest page tables, which defines the current address space in terms of guest-physical memory. This is under the exclusive control of the guest operating system. Furthermore, KVM does not need to keep track of changes to the guest page tables nor to context switches. In fact, assignments to `cr3` in non-root mode do not need to cause a `#vmexit` when the feature is enabled.

- In the x86 architecture, the format of the regular page tables (as pointed by the `cr3` register) are different from those of the nested pages. The KVM kernel module is therefore responsible for managing nested page tables, rooted at the `eptp` register on Intel architectures.

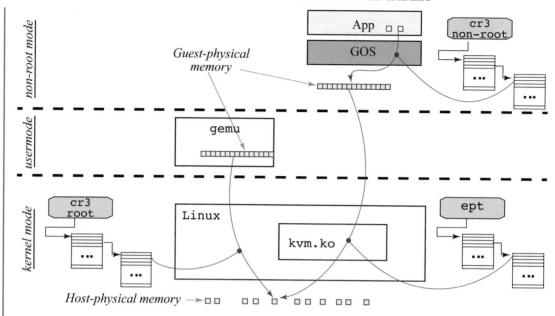

Figure 5.2: Access to guest-physical memory in KVM by the userspace QEMU/KVM component, the kvm.ko kernel module, and the guest operating system. Each red dot represents the entity that manages the corresponding hardware page table.

We make one more additional observation, which has performance implications: it is easier for the hypervisor to access guest-physical address space than it is to access guest-virtual address space. Specifically, the VT-x architecture lacks architectural support for `root mode` software to access efficiently the virtual address space of `non-root mode` environments.

Figure 5.2 shows that it is simple for the hypervisor to access the guest-physical address space: the userspace process can simply add a constant offset to reference a memory location. The KVM module itself can actually use the same approach, since the userspace process is already mapping the address space. It is, however, much harder and complex for the hypervisor to access the current virtual address spaces of the virtual machine as those mappings are only present in the MMU of the processor while the virtual machine is executing, but not when the hypervisor is executing.

Unfortunately, the guest instruction pointer is a virtual address and the KVM decoder must read the content of the faulting instruction from memory. Similarly, all the memory operands of the instruction refer to virtual addresses of the virtual machine. As a consequence, the KVM decoder and emulate makes repeated references in software to the guest page tables to determine the location in guest-physical memory of the instructions and operands.

The type-2 hypervisor design of KVM introduces a complication when managing guest-physical memory. Although the format differs, the semantic mapping of the guest-physical address space in the nested page tables must be consistent with the mappings of guest-physical address space of the host operating system. For example, should the guest-physical mapping change, e.g., a new page is allocated when handling an #vmexit from the virtual machine, the same mapping must also be reflected in address space of the QEMU process on the host. Conversely, should the host operating system decide to swap out a page from the QEMU process, the corresponding guest-physical extended mapping must also be removed.

The current version of KVM contains some key optimizations in the area of memory management. In particular, Linux's KSM mechanism [19] allows for the transparent sharing of memory, similar to the solution introduced in VMware ESX Server [177]. KSM allows memory pages with identical content to be transparently shared between Linux processes, and therefore between KVM virtual machines. A match is detected when two distinct pages belonging to potentially different processes, are found to have the same content and that content is declared stable, i.e., unlikely to change in the near future. KSM then (i) selects one page from the set; (ii) establishes read-only mappings in all address spaces to that page; and (iii) notifies the KVM module to do the same for the nested page tables; and (iv) deallocates all other replicas of the same page.

5.3   PERFORMANCE CONSIDERATIONS

Extended page tables provide significant performance benefits over shadow paging. It eliminates the need for memory tracing, which accounts for as much as 90% of all #vmexits in typical virtualized workloads [5]. It also eliminates the memory footprint of the shadow page tables.

However, extended paging is not free. A two-dimensional page-walk, as defined architecturally, is prohibitive in 64-bit address spaces: with 4 KB pages, the guest and the host page tables are each 4-level trees with $n = m = 4$ for a total of 24 architecturally defined memory references for a single TLB miss (vs. only 4 for a non-virtualized configuration). Fortunately, the existence of page walk caches inside the processors can limit the impact of the design [35].

Despite the ongoing focus on micro-architectural improvements, the overheads of virtualizing can be readily measured. Drepper measured it in 2008 [67], shortly after the introduction of the technology, using a simple pointer-chasing benchmark. Drepper's measurements show that the virtualized workload executes up to 17% slower on Intel CPU and up to 39% slower on AMD CPU when compared to the non-virtualized execution on the same processor. With such a benchmark, the difference in execution time is attributed nearly entirely to the difference in the TLB miss handling logic of the processors, and the number of memory references that are required to be fetched.

We perform a similar pointer-chasing experiment on a more recent, 2011-era Intel "Sandy Bridge" processor running at 2.4 Ghz. To measure the cost of TLB misses, we create a list that randomly links objects from a 32 GB heap, and then measure the average cost of a pointer reference. This micro-benchmark is designed to have no locality and to ensure that the great majority of pointer references will incur a TLB miss, irrespective of page size. Indeed, this benchmark has a TLB miss rate between 93% and 99.9%, depending on the configuration. The test only uses local memory for both data and page tables, as to avoid NUMA effects. Finally, the linked list is constructed to use different caches lines of the various pages, as to ensure that the last-level cache is efficiently used.

Figure 5.3 shows the performance of the different combination of (guest) page sizes and extended page sizes. Since the regular page table and the extended page table are orthogonal to each other, one can be larger or smaller than the other without any particular complication. The first block of results shows the performance for the standard 4 KB page size, with different extended page sizes, as well as for the non-virtualized case. The second and third blocks show the benefit of using super pages in the (guest) operating system. We report the performance in cycles per memory reference, which includes the cost of the memory reference (and the potential cache miss) as well as the cost of the TLB miss.

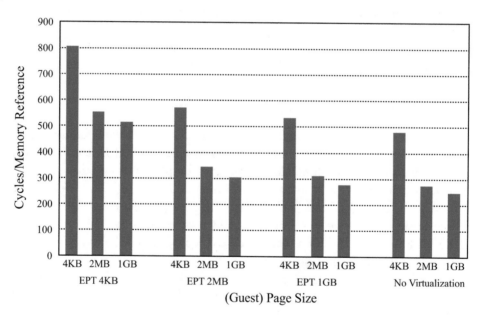

Figure 5.3: Measured cost of TLB misses on Intel "Sandy Bridge" processors for various combination of page sizes and extended page sizes.

In Figure 5.3, we first observe that the results of this micro-benchmark are all very high, with a reference costing at least 200 cycles on average. Fortunately, realistic workloads have much better TLB behavior because of the nearly universal locality principle [64]. We also note a substantial difference in the cost of these TLB misses: TLB misses of the most expensive configuration (4 KB pages, 4 KB extended pages) is 3.25× more expensive than the cheapest configuration (1 GB pages, non-virtualized). Finally, we observe that 4 KB page sizes are noticeably more expensive than 2 MB and 1 GB page sizes, but that 2 MB and 1 GB page lead to comparable TLB load times. Larger page sizes reduce the number of architecturally defined memory references required to load the TLB. For example, the number of required references goes from 24 down to 17 by simply increasing the extended page table size to 2 MB (since $m : 4 \rightarrow 3$). More importantly, the reduced number of page table entries dramatically reduces the probability that the memory references will not be served by the processor L3 cache.

Developers have recognized the need to carefully trade-off flexibility and performance in the management of memory of virtual machines. Today, 2 MB super pages available are a standard feature of commodity operating systems such as Linux, and most carefully tuned applications will take advantage of the feature. Many hypervisors today use 2 MB pages for extended page tables whenever possible. For example, VMware ESX preferably uses 2 MB extended pages. It uses 4 KB pages only when the system is under global memory pressure and the ballooning or transparent memory sharing mechanisms [177] kick in to reduce memory pressure [176].

5.4 FURTHER READING

Bhargava et al. [35] described the benefit of the **page walk cache** implemented in AMD Opteron. Ahn et al. [8] propose to use a flattened host page table array to reduce the memory references from $n \times m + n + m$ to $2n + 1$ references , with $n \leq 4$ based on the guest page size used. Gandhi et al. [72] recently measured the cost of nested paging on Intel "Sandy Bridge" CPU for large-memory workloads, and proposed to use segmentation instead of the conventional page table tree layout. Yaniv et al. propose to hash extended page table rather than the conventional 2D radix walk [189].

Agesen et al. [4] describe how introduction of extended paging and the dramatic reduction in the number of #vmexit, improves efficiency and allowed VMware to rely on direct execution to run most of the guest kernel code. More recently, Agesen et al. [5] describe how the judicious use of DBT for very short code sequences can be used to eliminate #vmexit and further increase performance.

CHAPTER 6

x86-64: I/O Virtualization

The previous chapters define virtual machines in terms of the three key attributes proposed by Popek and Goldberg—equivalence, safety, and performance—which help us to reason about virtualization from a CPU and MMU perspective. When introducing I/O capabilities to virtual machines, a fourth attribute becomes handy: **interposition**. The ability to interpose on the I/O of guest virtual machines allows the host to transparently observe, control, and manipulate this I/O, thereby decoupling it from the underlying physical I/O devices and enabling several appealing benefits.

The I/O that is generated and consumed by virtual machines is denoted **virtual I/O**, as opposed to **physical I/O**, which is generated and consumed by the operating system that controls the physical hardware. §6.1 enumerates the benefits of having interposable virtual I/O. §6.2 describes how physical I/O is conducted and provides the necessary background for the chapter. §6.3 describes how equivalent I/O capabilities are provided to guest VMs, without utilizing hardware support for I/O. §6.4 describes how to enhance the performance of virtual I/O with the help of hardware support that was added specifically for this purpose; we shall see that there is a cost to pay for this performance enhancement—losing the ability to interpose and all the associated benefits.

We note that, in this chapter, the notion of "hardware support" is additive to the previous chapters. Namely, unless otherwise stated, we assume that CPU and MMU virtualization is already supported by hardware and focus on hardware features that were added to help accelerate the processing of virtual I/O. Some parts of this chapter are independent of the specific architecture of the CPU, but other parts are architecture-dependent. In the latter case, as the title of this chapter suggests, we assume an Intel x86-64 processor with the VT-x virtualization extension [171]. In particular, with respect to architecture-dependent issues, this chapter describes Intel's x86-64 hardware support for I/O virtualization, which is called VT-d [1, 106].

6.1 BENEFITS OF I/O INTERPOSITION

The traditional, classic I/O virtualization approach—which predates hardware support—decouples virtual from physical I/O activity through a software indirection layer. The host exposes virtual I/O devices to its guests. It then intercepts (traps) VM requests directed at the virtual devices, and it fulfills (emulates) them using the physical hardware. This trap-and-emulate approach [78, 143] allows the host to interpose on all the I/O activity of its guests, as schematically illustrated in Figure 6.1.

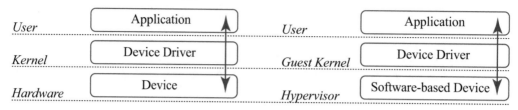

Figure 6.1: Traditional I/O virtualization decouples virtual I/O activity from physical I/O activity, allowing the hypervisor to interpose on the I/O of its virtual machines.

The benefits of I/O interposition are substantial [178]. First, interposition allows the hypervisor to encapsulate the entire state of the virtual machine—including that of its I/O devices—at any given time. The hypervisor is able to encode the state of the devices because (1) it implements these devices in software, while (2) interposing on each and every VM I/O operation. Encapsulation makes it possible for the hypervisor to, for example, suspend the execution of a VM, store its encoded representation (denoted as the VM image), and resume execution at a later time. The combination of encapsulation and decoupling of virtual from physical I/O is especially powerful, providing transparent VM portability between different servers that may be equipped with different I/O devices. The hypervisor's role is to map between virtual to physical I/O devices. Thus, it can suspend the VM's execution on a source server, copy its image to a target server, and resume execution there [115], even if the physical I/O devices at the source and target servers are different. All that is required is for the target hypervisor to recouple the virtual devices of the VM with the locally available physical devices. The VM remains unaware throughout the move—nothing has changed from its perspective. In fact, the VM need not stop execution while it moves between servers, a feature known as **live migration** [51, 135].

The hypervisor can exploit its ability to interpose on the I/O and dynamically decouple/recouple a physical device from/to a VM even if the VM stays on its original machine rather than moves. For example, when upgrading, reconfiguring, or otherwise modifying the VM's storage device, the hypervisor can use its indirection layer to hide the modifications, e.g., by copying the stored content from an old device to a new one, in a transparent manner, while the VM is running.

Arguably, the most notable benefit of CPU virtualization is consolidation, namely, the ability to run multiple virtual servers on a much smaller number of physical servers, by multiplexing the hardware. With classic I/O virtualization, it is the ability to interpose on the I/O that provides I/O device consolidation. Specifically, interposition allows the hypervisor to map multiple virtual devices onto a much smaller set of physical devices, thereby increasing utilization, improving efficiency, and reducing costs.

I/O interposition additionally allows the hypervisor to aggregate multiple physical I/O devices into a single, superior virtual device. The virtual device provides better performance and robustness by seamlessly unifying physical devices, exposing them to VMs as one, load balancing between them, and hiding failures of individual devices by transparently falling back on other devices.

I/O interposition further allows the hypervisor to add support for new features that are not natively provided by the physical device. Examples include: replicated disk writes to transparently recover from disk failures; preserving old versions of disk blocks, instead of overwriting them, to allow for snapshots, record-replay, and time travel capabilities; storage deduplication, which eliminates duplicate copies of repeated data and thus utilizes a given storage capacity more effectively; seamless networking and storage compression/decompression and encryption/decryption; metering, accounting, billing, and rate-limiting of I/O activity; quality of service guarantees and assurances; and security-related features such network intrusion detection and deep packet inspection.

Lastly, while not immediately obvious, traditional I/O virtualization realized through I/O interposition allows hypervisors to apply all the canonical memory optimizations to the memory images of VMs, including memory overcommitment (where the combined size of the virtual address spaces allocated to VMs exceeds the size of the physical memory), demand-based paging and lazy allocation based on actual memory use, swapping of unused memory to disk, copy-on-write, page migration, transparent huge pages, and so on. If VMs were allowed to use physical I/O devices directly, these optimizations would not have been possible, because (1) I/O devices are often capable of directly accessing the memory on their own without CPU involvement, (2) this type of direct access does not tolerate page faults, and (3) hypervisors have no general way to know which memory regions would serve as target locations of such direct device accesses. We further discuss this issue below.

> In summary, classic I/O virtualization is implemented via I/O interposition, which provides many advantages such as added features and state encapsulation that allows for migration.

6.2   PHYSICAL I/O

The previous section conducted a high-level discussion about the advantages of traditional virtual I/O. Next, our goal is to describe how virtual I/O really works. In order to do that, we first need a better understanding of physical I/O, as virtual machines believe that they perform such I/O, and it is the hypervisor's job to maintain this illusion. A lot of topics, details, and legacy issues fall under the title of physical I/O that hypervisors need to virtualize—much too many to fit in the limited space of this book. Therefore, in this section, we only briefly describe some key mechanisms that are required to allow us to explain the more important aspects of virtual I/O; these mechanisms would then be used in the following sections. Readers who already have reasonable understanding regarding how physical I/O works, notably over PCIe fabrics, may skip this section.

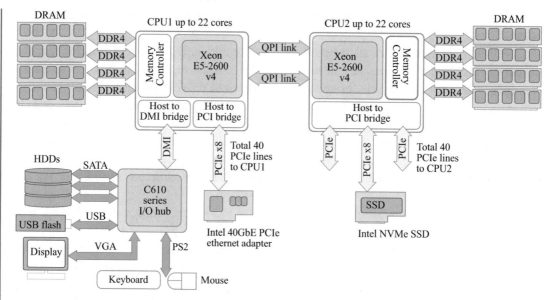

Figure 6.2: High-level view of a possible internal organization of an Intel server. In this example, the server houses two CPUs from the E5-2600 v4 family, each of which may consist of 4–22 cores, depending on the specific model.

6.2.1   DISCOVERING AND INTERACTING WITH I/O DEVICES

A typical computer consists of a wide variety of I/O devices, including network controllers, disk drives, video, mouse, and keyboard. Figure 6.2 depicts a high-level view of a possible organization of a modern Intel server. Such a system may house several CPU packages, which communicate via the Intel QuickPath Interconnect (QPI) [134]. CPU cores access the memory modules (DIMMs) through their memory controllers, and they communicate with the various I/O devices through the relevant host bridges. Notably, communications with Peripheral Component Interconnect Express (PCIe) devices flow through the host-to-PCIe bridge [141]. The PCIe fabric poses the greatest challenge for efficient virtualization, because it delivers the highest throughput as compared to other local I/O fabrics. For this reason, hardware support for I/O virtualization largely focuses on PCIe; this chapter likewise focuses on virtualizing PCIe devices, for the same reason.

Upon startup, at boot time, the operating system kernel must somehow figure out which devices are available in the physical machine. The exact way to obtain this goal is nontrivial; it varies and depends on several factors, but typically it involves the Basic Input/Output System (BIOS) firmware [63], or its successor, the Unified Extensible Firmware Interface (UEFI) [170]. The firmware (BIOS or UEFI) then provides a description of the available devices in some standard format, such as the one dictated by the Advanced Configuration and Power Interface (ACPI) [169]. In §6.2.3, we provide a detailed example of how PCIe devices are discovered. Generally speaking, however, the device discovery process requires the host OS to query the physical firmware/hardware. Thus, when virtualizing the OS, it will be the hypervisor that will be queried, allowing it to present devices as it sees fit to the guest OS, irrespective of the real physical devices.

Let us assume that the I/O devices have been discovered. In this state, they interact with the CPU and the memory in three ways, as depicted in Figure 6.3 and as discussed next.

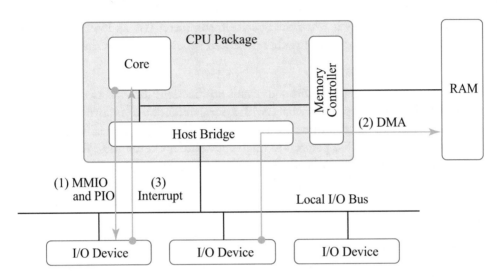

Figure 6.3: There are three ways that I/O devices can interact with the CPU and the memory: (1) MMIO and PIO allow CPU cores to communicate with I/O devices; (2) DMA allows I/O devices to access the memory; and (3) interrupts allow I/O devices to communicate with CPU cores.

PIO and MMIO: Port-mapped I/O (PIO) and memory-mapped I/O (MMIO) provide the most basic method for CPUs to interact with I/O devices. The BIOS or UEFI associate the registers of the I/O devices with unique, dedicated addresses. The CPU often uses these addresses to implement control channels, namely to send commands to the I/O devices and then to receive the corresponding statuses by polling the appropriate registers for reading.

Addresses of PIO—called "ports"—are separated from the memory address space and have their own dedicated physical bus. Such addresses are typically limited to 16 bits. They are used via the special OUT and IN x86 instructions, which write/read 1–4 bytes to/from the I/O devices. The addresses are usually well known and set by hardware vendors; for example, ports 0x0060–0x0064 are used to read from and send commands to keyboards and mice with PS/2 connectors.

MMIO is similar, but the device registers are associated with physical memory addresses, and they are referred to using regular load and store x86 operations through the memory bus. The association between device registers and memory addresses is predetermined on startup. The host bridge controller and memory controller are aware of these associations and route data accordingly.

DMA: Using PIO and MMIO to move large amounts of data from I/O devices to memory and vice versa can be highly inefficient. Such data transfers might take a long time, during which the core must be continuously and synchronously involved to perform explicit in/out/load/store instructions. A better, more performant alternative is to allow I/O devices to access the memory directly, without CPU core involvement. Such interaction is made possible with the **Direct Memory Access** (DMA) mechanism. The core only initiates the DMA operation, asking the I/O device to asynchronously notify it when the operation completes (via an interrupt, as discussed next). The core is then free to engage in other unrelated activities between the DMA initiation and completion.

Interrupts: I/O devices trigger asynchronous event notifications directed at the CPU cores by issuing interrupts. Each interrupt is associated with a number (denoted "interrupt vector"), which corresponds to an entry in the x86 Interrupt Descriptor Table (IDT). The IDT is populated on startup by the OS with (up to) 256 pointers to interrupt-handler OS routines. When an interrupt fires, the hardware invokes the associated routine on the target core, using the interrupt vector to index the IDT. The OS declares the location and size of the IDT in a per-core IDT register (IDTR). The IDTR value can be different or the same across cores; it holds a virtual address.

LAPIC: During execution, the OS occasionally performs interrupt-related operations, including: enabling and disabling interrupts; notifying the hardware upon interrupt handling completion; sending inter-processor interrupts (IPIs) between cores; and configuring the timer to deliver clock interrupts. All these operations are performed through the per-core Local Advanced Programmable Interrupt Controller (LAPIC). The LAPIC interrupt request register (IRR) constitutes a 256-bit read-only bitmap marking fired interrupt requests that have not yet been handled by the core. The LAPIC in-service register (ISR) similarly marks interrupts that are currently being handled. When the OS enables interrupt delivery, the LAPIC clears the highest-priority bit in the IRR, sets it in the ISR, and invokes the corresponding handler. The OS signals the completion of the handling of the interrupt via the end-of-interrupt (EOI) LAPIC register, which clears the highest-priority bit in the ISR. Interrupts are sent to other cores using the interrupt command register (ICR). Registers of the newest LAPIC interface, x2APIC, are accessed through read and write operations of model-specific registers (MSRs), as opposed to previous interfaces (such as xAPIC) that were accessed via regular load/store operations from/to a predefined MMIO area.

6.2.2   DRIVING DEVICES THROUGH RING BUFFERS

I/O devices, such as PCIe solid state drives (SSDs) and network controller (NICs), can deliver high throughput rates. For example, as of this writing, devices that deliver 10–100 Gbit/s are a commodity. To understand how such high-throughput devices are virtualized, it would be helpful to understand how they work. Overwhelmingly, these devices stream their I/O through one or more producer/consumer **ring buffers**. A ring is a memory array shared between the OS device driver and the associated device, as illustrated in Figure 6.4. The ring is circular in that the device

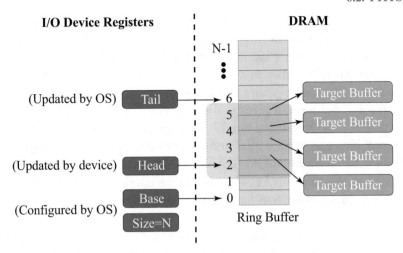

Figure 6.4: High-throughput I/O devices typically use a ring buffer to communicate with their OS device driver. The shaded area marks the range of DMA descriptors currently accessible to the device. Each entry points to a DMA target buffer. The device will write incoming data into DMA target buffers or read outgoing data from DMA target buffers, depending on the direction of the DMA.

and driver wrap around to the beginning of the array when they reach its end. The entries in the ring are called DMA descriptors. Their exact format and content may vary between I/O devices, but they typically specify at least the address and size of the corresponding DMA target buffers—the memory areas used by the device DMAs to write/read incoming/outgoing data. The descriptors also commonly contain status bits that help the driver and the device to synchronize.

Devices must also know the direction of each requested DMA, namely, whether the data should be transmitted from memory (into the device) or received (from the device) into memory. The direction can be specified in the descriptor, as is typical for disk drives, or the device can employ different rings for receive and transmit activity, as is typical for NICs. (In the latter case, the direction is implied by the ring.) The NIC receive and transmit rings are denoted Rx and Tx, respectively. NICs employ at least one Rx and one Tx per port (physical cable). They may employ multiple Rx/Tx rings per port to promote scalability, as different rings can be easily handled concurrently by different cores.

Upon initialization, the OS device driver allocates the rings and configures the I/O device with the ring sizes and base locations. For each ring, the device and driver utilize a head and a tail pointers to delimit the ring content that can be used by the device: [$head,tail$). The base, size, head, and tail registers of the device ring are accessible to the OS driver via MMIO. The device iteratively consumes (removes) descriptors from the head, and it increments the head to point to the next descriptor to be used subsequently. Similarly, the driver adds descriptors to the tail, incrementing the tail to point to the entry that it will use subsequently. The Rx ring is full during a period of I/O inactivity. The Tx ring is empty during a period of I/O inactivity. The device and driver may distinguish between ring emptiness and fullness—when head and tail point to the same location in Figure 6.4—using the aforementioned descriptor status bits; for example, "produced" and "consumed" bits that are set by the driver and the device, respectively.

To illustrate, assume the OS wants to transmit two packets after a period of network inactivity. Initially, the Tx head and tail point to the same descriptor k, signifying (with status bits) that Tx is empty. The OS driver sets the k and $k + 1$ descriptors to point to the two packets, turns on their "produced" bits, and lets the NIC know that new packets are pending by updating the tail register to point to $k + 2$ (modulo N). The NIC sequentially processes the packets, beginning at the head (k), which is incremented until it reaches the tail ($k + 2$). With Tx, the head always "chases" the tail throughout the execution, meaning the NIC tries to send the packets as fast as it can.

The device asynchronously informs its OS driver that data was transmitted or received by triggering an interrupt. The device coalesces interrupts when their rate is high. Upon receiving an interrupt, the driver of a high-throughput device handles the I/O burst. Namely, it sequentially iterates through and processes all the descriptors whose corresponding DMAs have completed. In the case of Rx, for example, processing includes handing the received packets to the TCP/IP stack and rearming the Rx ring with new buffers.

6.2.3   PCIE

As noted above, much of the advancements in hardware support for I/O virtualization revolve around PCI Express (Peripheral Component Interconnect Express; officially abbreviated as PCIe) [141], which is a specification of a local serial bus that is standardized by the PCI Special Interest Group (PCI-SIG) industry consortium. Architecturally, PCIe is similar to a lossless network infrastructure: it consists of a layers (transaction, data link, physical); it transmits packets over links from one node to another according to well-defined routing rules; and it has flow control, error detection, retransmission, and quality of service.

Hierarchy: The topology of the PCIe fabric is arranged as a tree, as illustrated in Figure 6.5. (We will later see that hardware support for I/O virtualization dictates dynamic runtime changes in this tree.) The root of the tree is the host bridge (also shown in Figure 6.3), which channels all types of PCIe traffic to/from the cores/memory: PIO, MMIO, DMA, and interrupts. The host bridge resides in the Root Complex (RC), usually on the CPU. The edges of the tree are PCIe buses, and the nodes of the tree are PCIe functions. Functions are either bridges (inner nodes) or endpoints (leaves or root). Endpoints correspond to individual I/O channels in physical PCIe devices; for example, a dual-port NIC has two endpoints—one per port. A bridge connects two buses, and a switch aggregates two or more bridges. A bus connects one upstream node to at most 32 downstream devices, such that each device houses up to 8 PCIe functions. Each bus consists of 1–32 PCIe lanes (unrelated to the hierarchy); in the current version of PCIe (v3), each lane delivers 985 MB/s, both upstream (toward the RC) and downstream (toward the endpoint).

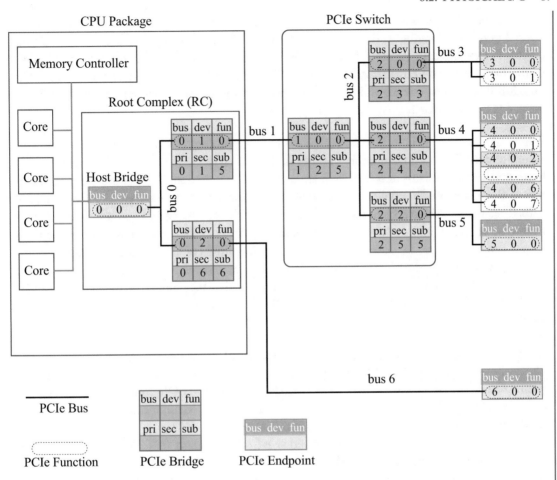

Figure 6.5: PCIe tree topology example. The abbreviations "dev", and "fun" stand for device and function. The abbreviations "pri", "sec", and "sub" stand for primary, secondary, and subordinate.

Node Enumeration: Every PCIe function (node) and PCIe bus (edge) in the PCIe graph is uniquely identified. Let N be a PCIe node in the tree. N's identifier is a 16-bit number in the form of `bus:device.function` (BDF), such that `bus`, `device`, and `function` consist of 8, 5, and 3 bits.[1] As can be seen in Figure 6.5, N's `bus` is the number of the bus that directly connects to N from upstream—let us denote this bus as B. N's `device` is a serial number belonging to the

---

[1]Be warned of the overloaded terms. A `function` number is different than a PCIe function. The former is a 3-bit index, whereas the latter is physical hardware, as well as the type of the node of the PCIe tree identified by a full BDF. The same applies to the `device` 5-bit number, which is different than a physical device, which serves as an aggregator of PCIe functions.

enumeration of all the devices that are immediately reachable through B. Finally, N's `function` is a serial number belonging to the enumeration of all the functions housed by N's encapsulating device. For example, in Figure 6.5, `3:0.0` (top/right) is an endpoint PCIe function that resides in the first (and only) downstream device reachable through Bus 3; `3:0.1` is the second PCIe function in the same device; and `2:2.0` is a bridge PCIe function residing in the third device reachable through Bus 2.

Edge Enumeration: Figure 6.5 shows that the PCIe graph edges (buses) are uniquely numbered from 0 and up. The maximal number of buses in the system is 256 (as `bus` has 8 bits). Recursively, buses that are descendants of some bus B are numerically larger than B, and buses that are ancestors of B are numerically smaller. This enumeration allows every PCIe bridge G to be associated with three buses: primary, secondary, and subordinate. The primary and secondary are, respectively, the upstream and downstream buses that G connects. G's subordinate is its numerically largest descendant bus. In follows that the numeric identifiers of all the buses that reside under G range from its secondary bus to its subordinate bus. For example, the buses under bridge `1:0.0` range from 2 (secondary) to 5 (subordinate). Importantly, the primary/secondary/subordinate per-bridge attributes accurately describe the PCIe hierarchy and determine how PCIe packets are routed.

Configuration Space: As noted in §6.2.1, on startup, the OS typically discovers the system's I/O devices by using the BIOS/UEFI firmware, which provides device information via some standard format like ACPI tables. For PCIe devices, discovery involves the PCIe configuration space array, as depicted in Figure 6.6. This array is accessible to the OS via MMIO operations. Later on, when we virtualize the OS, we will need to simulate this array for the VM.

The OS can find the array's address and size in the standard "MCFG" APCI table (Figure 6.6a).[2] In our example, MCFG indicates that the array should include an entry for every valid BDF. Since there are $(2^{16} =)$ 64 KB such BDFs, and since PCIe specifies that the per-BDF configuration space size is 4 KB, then the array size is (64 KB × 4 KB =) 256 MB. The configuration space associated with a given BDF can be found by adding 4 KB × BDF to the address of the array.

The 4 KB configuration space of a PCIe node identified by BDF consists of three parts (Figure 6.6c). The first 256 bytes constitute a valid PCI (rather than PCIe) configuration space, for backward compatibility (PCIe superseded PCI). Of these, the first 64 bytes constitutes the header (Figure 6.6d), which identifies such attributes as the functional class of the device (network, storage, etc.), the vendor ID (assigned by PCI-SIG), and the device ID (assigned by the vendor). These attributes allow the OS to identify the devices, associate them with appropriate drivers, initialize them, and make them available for general use.

---

[2]Running the "dmesg | grep MMCONFIG" shell command in Linux prints the line that contains the string MMCONFIG in the output of the OS boot process. (MMCONFIG stands for "memory-mapped configuration space"). The result is: `PCI: MMCONFIG for domain 0000 [bus 00-ff] at [mem 0xe0000000-0xefffffff]`, in accordance to Figure 6.6b.

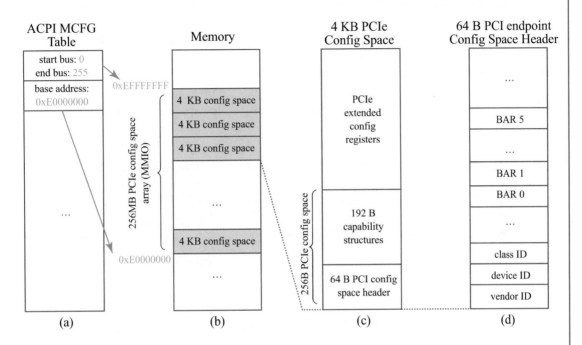

Figure 6.6: The PCIe configuration space on x86 is accessible via the MCFG APCI table. Subfigure (d) depicts the header of an endpoint. The header of a bridge is somewhat different; notably, it specifies the bridge's primary, secondary, and subordinate buses, thereby allowing the OS to reconstruct the PCIe tree.

Each endpoint can have up to six Base Address Registers (BARs), which publicize the MMIO (or PIO) addresses to be used when the OS wishes to interact with the device—the locations where device registers are found. For example, in the device depicted in Figure 6.4, corresponding configuration space BARs will specify where to find the head, tail, base, and size registers. The exact semantics of the areas pointed to by the BARs are determined by the manufacturers of the devices.

MSI: So far, we considered two address space types for I/O: port-mapped (PIO) and memory-mapped (MMIO). PCIe supports a third type—for interrupts. Message Signaled Interrupts (MSI) allow a device to send a PCIe packet whose destination is a LAPIC of a core; the third address space is thus populated with addresses of all the system's LAPICs. An MSI interrupt is similar to a DMA operation, but instead of targeting the memory, it targets a LAPIC. The MSI interrupt propagates upstream through the PCIe hierarchy until it reaches the host bridge (Figure 6.5), which forwards it to the destination LAPIC that is encoded in the packet.

The OS configures a device to use MSI interrupts by writing the address of the target LAPIC and the desired interrupt vector to the Message-Address and Message-Data registers, which reside in the capability structures area of the configuration space (middle part of Figure 6.6c). When firing an interrupt, the device then sends the content of Message-Data to Message-Address. MSI supports up to 32 interrupts per device, whereas the newer MSI-X supports up to 2048.[3]

We remark that the OS configures the PCIe configuration space using the usual PIO/MMIO mechanisms—from the OS perspective, the configuration space is really just another I/O device.

> In summary, OSes interact with I/O devices via PIO, MMIO, DMA, and interrupts mediated through LAPICs. OS drivers usually drive high-throughput I/O devices through ring buffers, which reside in MMIO space. PCIe devices are arranged in a BDF tree hierarchy. The PCIe configuration space encodes this hierarchy, while providing (MMIO) pointers to device registers.

6.3   VIRTUAL I/O WITHOUT HARDWARE SUPPORT

A guest operating system does not cease to behave as an operating system merely because it runs as a guest. Generally speaking, it still believes that it exclusively controls all the physical I/O devices, and it discovers, initializes, and drives these devices exactly as described in §6.2. The problem is that, without explicit hardware support, the hypervisor typically cannot allow the guest to interact with I/O devices in this way. Consider, for example, the system's hard drive controller. This device must be safely shared between the hypervisor and its guests in a manner that allows all of them to peacefully co-exist. But our guest is unaware of the fact that it must share, and it would not know how even if it did. Consequently, allowing the guest to access the disk drive directly would most likely result in an immediate crash and permanent data loss. A similar case can be made for other I/O devices.

To avoid this problem, the hypervisor must prevent guests from accessing real devices while sustaining the illusion that devices can be accessed; the hypervisor must therefore "fake" I/O devices for its guests, denoted as virtual I/O devices. The hypervisor achieves this goal by trapping all the guest's I/O-related operations and by emulating them to achieve the desired effect.

---

[3]MSI/MSI-X replaced the legacy IOAPIC: an interrupt controller for the entire CPU package that interacted with individual LAPICs. IOAPIC supported only 24 interrupts, it sometimes violated DMA/interrupt ordering, and it required dedicated wiring.

6.3.1   I/O EMULATION (FULL VIRTUALIZATION)

In §1.5, we briefly mentioned the concept of I/O emulation, which may also be called *full* I/O virtualization. Here we describe in more detail how this concept is realized. In §6.2.1, we have noted that (1) the OS discovers and "talks" to I/O devices by using MMIO and PIO operations, and that (2) the I/O devices respond by triggering interrupts and by reading/writing data to/from memory via DMAs. Importantly, no other type of OS/device interaction exists. The hypervisor can therefore support the illusion that the guest controls the devices by (1) arranging things such that every guest's PIO and MMIO will trap into the hypervisor, and by (2) responding to these PIOs and MMIOs as real devices would: injecting interrupts to the guest and reading/writing to/from its (guest-physical) memory as if performing DMAs.

Emulating DMAs to/from guest memory is trivial for the hypervisor, because it can read from and write to this memory as it pleases. Guest's MMIOs are regular loads/stores from/to guest memory pages, so the hypervisor can arrange for these memory accesses to trap by mapping the pages as reserved/non-present (both loads and stores trigger exits) or as read-only (only stores trigger exits). Guest's PIOs are privileged instructions, and the hypervisor can configure the guest's VMCS to trap upon them. Likewise, the hypervisor can use the VMCS to inject interrupts to the guest. (Before VT-x and VMCSes: the hypervisor injected interrupts by directly invoking the appropriate interrupt handler routine pointed to by the guest's IDT, which was emulated by the hypervisor; PIOs were replaced by hypercalls using dynamic binary translation; and MMIOs and DMAs were emulated identically to how they are emulated with VT-x, as described above.)

Figure 6.7 schematically illustrates the aforementioned interactions for the KVM/QEMU hypervisor. Generally, every hosted virtual machine is encapsulated within a QEMU process, such that different VMs reside in different processes. Internally, a QEMU process represents the VCPUs (virtual cores) of its VM using different threads. Additionally, for every virtual device that QEMU hands to its VM, it spawns another thread, denoted as "I/O thread". VCPU threads have two execution contexts: one for the guest VM and one for the host QEMU. The role of the host VCPU context is to handle exits of the guest VCPU context. The role of the I/O thread is to handle asynchronous activity related to the corresponding virtual device, which is not synchronously initiated by guest VCPU contexts. For example, handling incoming network packets.

The illustration in Figure 6.7 depicts a VM that has two VCPUs and one I/O device. The guest VM device driver issues MMIOs/PIOs to drive the device. But the device is virtual, so these operations are directed at ordinary, read/write-protected memory locations, triggering exits that suspend the VM VCPU context and invoke KVM. The latter relays the events back to the very same VCPU thread, but to its host, rather than guest, execution context. QEMU's device emulation layer then processes these events, using the physical resources of the system, typically through regular system calls. The emulation layer emulates DMAs by writing/reading to/from the guest's I/O buffers, which are accessible to it via shared memory. It then resumes the guest execution context via KVM, possibly injecting an interrupt to signal to the guest that I/O events occurred.

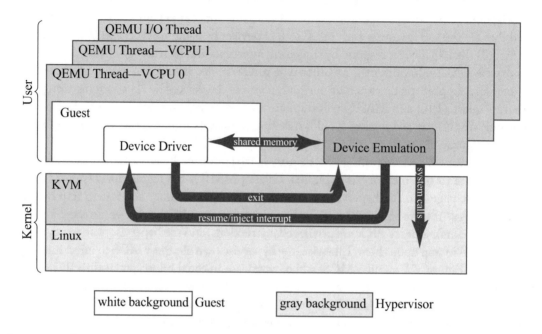

Figure 6.7: I/O emulation in the KVM/QEMU hypervisor. The guest device driver and QEMU's device emulation are operational in all VCPU threads (shown only in VCPU 0). The host execution context of the VCPU handles synchronous exits of the guest execution context. The I/O thread handles asynchronous activity. The shared memory channel corresponds to buffers that the guest posts for DMA.

By utilizing the emulation technique we just described, the hypervisor can decide which set of virtual I/O devices to expose to the guest. Figure 6.8 lists the set of PCI devices exposed to a typical Linux virtual machine hosted by QEMU/KVM. The figure shows the output of the lspci shell utility when invoked inside the VM (lspci prints the configuration space hierarchy depicted in Figure 6.6). Different than a bare-metal OS, the guest's lspci does not read the real ACPI tables and configuration space that were established by the real BIOS/UEFI firmware. Instead, it reads the content of emulated ACPI tables and configuration space, as generated by an emulated BIOS/UEFI. We can see that, by default, QEMU emulates the Intel 440FX host bridge controller as the root of the BDF hierarchy (00:0.0). In this hierarchy, we can further see that QEMU emulates two NICs for this VM: Intel's 82540EM Gigabit Ethernet Controller (00:03.0), and Red Hat's Virtio network device (00:04.0). Next, we discuss the emulated Intel NIC, to give one concrete example of how I/O emulation works for a specific device. We defer discussion about the other NIC to §6.3.2, where we define, motivate, and explain I/O paravirtualization.

```
00:00.0 Host bridge: Intel Corporation 440FX - 82441FX PMC [Natoma] (rev 02)
00:01.0 ISA bridge: Intel Corporation 82371SB PIIX3 ISA [Natoma/Triton II]
00:01.1 IDE interface: Intel Corporation 82371SB PIIX3 IDE [Natoma/Triton II]
00:01.3 Bridge: Intel Corporation 82371AB/EB/MB PIIX4 ACPI (rev 03)
00:02.0 VGA compatible controller: Device 1234:1111 (rev 02)
00:03.0 Ethernet controller: Intel Corporation 82540EM Gigabit Ethernet
00:04.0 Ethernet controller: Red Hat, Inc Virtio network device
00:05.0 SCSI storage controller: Red Hat, Inc Virtio block device
```

Figure 6.8: Output of the lspci shell command executed within a typical Linux VM, which is running on top of the KVM/QEMU hypervisor.

```
00:03.0 Ethernet controller: Intel 82540EM Gigabit Ethernet Controller
        Flags: bus master, fast devsel, latency 0, IRQ 11
        Memory at febc0000 (32-bit, non-prefetchable) [size=128K]
        I/O ports at c000 [size=64]
        Expansion ROM at feb40000 [disabled] [size=256K]
        Kernel driver in use: e1000
```

Figure 6.9: Partial lspci output for the emulated Intel NIC, identifying its guest driver as e1000, and revealing its MMIO BAR addresses (0xfebc0000) as specified in its emulated configuration space.

e1000:   The 82540EM NIC corresponds to an old network controller that was launched by Intel in 2002 [100] and has since been superseded by multiple generations of newer models. Naturally, our physical host machine does not have this NIC (it is equipped with a newer model by a different vendor). The NIC's physical absence is not an issue, however, as it is emulated purely by software. In fact, the 82540EM is so old that it predates PCIe and only supports PCI. But this too is not a issue (regardless of PCIe being backward compatible with PCI), because QEMU exposes the BDF hierarchy as PCI in any case. That is, the emulated 00:00.0 Intel 440FX host bridge (root of BDF hierarchy) is a PCI- rather than PCIe-to-host bridge. Since the bridge is emulated by software, it does not matter that the underlying physical I/O system uses the newer PCIe fabric.

The 82540EM model of the emulated Intel NIC is not particularly special or important in any sense. What is important is for the VM to have an appropriate *driver* for the hypervisor to match with its own network emulation layer, such that these two software components would be compatible (the two rounded rectangles in Figure 6.7). Using the lspci utility, we can print detailed configuration space and other information of PCI devices. Figure 6.9 shows some of this information for our emulated Intel NIC. The last line indicates that the guest kernel driver associated with this NIC is e1000 [99]. The latter is the name of a legacy Intel driver that ships by default with all prevalent OSes, and has been shipping with them for years. As such, all prevalent hypervisors rely on the presence of e1000 in guest OSes to provide default networking capabilities, utilizing it as the de-facto standard for network emulation. The e1000 driver supports the family of Intel's PCI/PCI-X Gigabit Ethernet NICs (the 82540EM is simply a member of this family).

The third line in Figure 6.9 specifies the MMIO BAR address of the emulated NIC, where its registers are found. The semantics of the registers and their exact location relative to the BAR are determined by the specification of the NIC [101]. Table 6.1 lists some of the central registers used by the e1000 driver. The receive and transmit categories directly correspond to the base, size, head, and tail registers of the Rx and Tx rings, as described in §6.2.2. When the guest attempts to access a register, KVM suspends it and resumes QEMU host context, providing it with the register's address and a description of the attempted operation. QEMU then responds appropriately based on the NIC specification. By faithfully adhering to this specification, and by using the same BAR, the emulation layer behaves exactly as the guest e1000 driver expects and produces an outcome identical to a functionally-equivalent physical NIC.

Table 6.1: Key NIC registers [101] used by the e1000 driver [99]. When added to the MMIO BAR specified in the configuration space, the offset provides the address of the corresponding registers.

Category	Name	Abbreviates	Offset	Description
receive	RDBAH	receive descriptor base address	0x02800	base address of Rx ring
	RDLEN	receive descriptor length	0x02808	Rx ring size
	RDH	receive descriptor head	0x02810	pointer to head of Rx ring
	RDT	receive descriptor tail	0x02818	pointer to tail of Rx ring
transmit	TDBAH	transmit descriptor base address	0x03800	base address of Tx ring
	TDLEN	transmit descriptor length	0x03808	Tx ring size
	TDH	transmit descriptor head	0x03810	pointer to head of Tx ring
	TDT	transmit descriptor tail	0x03818	pointer to tail of Tx ring
other	STATUS	status	0x00008	current device status
	ICR	interrupt cause read	0x000C0	bitmap of causes
	IMS	interrupt mask set	0x000D0	enable interrupts
	IMC	interrupt mask clear	0x000D8	disable interrupts

The QEMU e1000 emulation layer (file hw/net/e1000.c in the QEMU codebase) implements the NIC registers by utilizing an array named mac_reg. According to the NIC specification, for example, the ICR register (which specifies the reason for the last interrupt) is cleared upon a read. Thus, when the guest reads the emulated ICR, the read operation triggers an exit, which transfers control to KVM, which resumes the host context of QEMU, which analyzes the instruction that triggered the exit and consequently invokes the mac_icr_read routine. The code of mac_icr_read is listed in Figure 6.10. As can be seen, the emulation layer correctly emulates the NIC's specification by first saving the ICR value in a local variable, zeroing the register, and only then returning the previous value.

6.3.2   I/O PARAVIRTUALIZATION

While I/O emulation implements a correct behavior, it might induce substantial performance overheads, because efficient emulation was not recognized as a desirable feature when the phys-

```
static uint32_t mac_icr_read(E1000State *s)
{
    uint32_t ret = s->mac_reg[ICR];
    s->mac_reg[ICR] = 0;
    return ret;
}
```

Figure 6.10: The QEMU routine that emulates guest e1000 ICR read operations (simplified version).

ical device was designed. It is probably safe to assume that the original designers of the Intel 82540EM Gigabit Ethernet Controller were not aware of the possibility that this device will be routinely virtualized, let alone that its interface would become the de-facto standard for network virtualization. Indeed, sending/receiving a single Ethernet frame via e1000 involves multiple register accesses, which translate to multiple exits per frame in virtualized setups.

There are other contributing factors to the overheads of emulation. For example, some read-only e1000 NIC registers are frequently accessed but do not have side-effects when they are being read. Consider the STATUS register, for example, which is accessed twice for every sent message. STATUS reports the current status of the NIC without changing the NIC's state. Thus, in principle, the hypervisor could have supported STATUS reads that do not trigger exits, by providing the VM with read-only permissions for this register. But this potential optimization is not possible. Because efficient virtualization was not a design goal, the NIC registers are tightly packed in the MMIO space, such that STATUS and ICR reside on the same memory page. STATUS must therefore be read-protected, because memory permissions work in page granularity, and ICR must be read-protected for correctness (Figure 6.10).

Virtualization overheads caused by inefficient interfaces of physical devices could, in principle, be eliminated, if we redesign the devices to have virtualization-friendlier interfaces. Such a redesign is likely impractical as far as *physical* devices are concerned. But it *is* practical for *virtual* devices, which are exclusively implemented by software. This observation underlies I/O paravirtualization, whereby guests and hosts agree upon a (virtual) device specification to be used for I/O emulation, with the explicit goal of minimizing overheads. Identically to baseline I/O emulation, the guest uses a device driver that is served by a matching host emulation layer, as depicted in Figure 6.7. The difference is that the specification of this virtual device is defined with virtualization in mind, so as to minimize the number of exits. Such a device is said to be paravirtual (rather than fully virtual), as it makes the guest "aware" that it is being virtualized: the guest must install a special device driver that is only compatible with its hypervisor, not with any real physical hardware.

I/O paravirtualization offers substantial performance improvements for some workloads as compared to I/O emulation. But there are also disadvantages. As noted, paravirtualization oftentimes requires that guest OS administrators install the paravirtual drivers. (Administrators of production-critical servers usually tend to prefer not to install new software if they can help it, fearing that the new software is somehow incompatible with the existing software.) Paravirtual-ization is additionally less portable than emulation, as drivers that work for one hypervisor will typically not work for another. Additionally, hypervisor developers who wish to make paravirtual-ization available for their guest OSes will need to implement and maintain a (different) paravirtual device driver for every type of guest OS they choose to support.

Virtio: The framework of paravirtual I/O devices of KVM/QEMU is called virtio [154, 168], offering a common guest-host interface and communication mechanism. Figure 6.11 shows some of the devices that are implemented using this framework, notably, network (virtio-net), block (virtio-blk), and character (virtio-console). As usual, each paravirtual device driver (top of the figure) corresponds to a matching emulation layer in the QEMU part of the hypervisor (bottom). Despite being paravirtual, virtio devices are exposed to guests like any other physical/emulated device and thus require no special treatment in terms of discovery and initialization. For example, in Figure 6.8, the BDF 00:04.0 is associated with a virtio-net NIC. The configuration space information of this NIC reveals all the data that is required in order to use this NIC, for example, the location of its Rx and Tx rings and their registers.

The central construct of virtio is virtqueue, which is essentially a ring (Figure 6.4) where buffers are posted by the guest to be consumed by the host. Emulated devices are inefficient largely because every MMIO/PIO access implicitly triggers an exit. In contrast, guests that use virtqueues *never* trigger exits unless they consciously intend to do so, by invoking the virtqueue_kick function, to explicitly let the hypervisor know that they require service. By minimizing the number of kicks, guests reduce exit overheads and increase batching.

In addition to the ring, each virtqueue is associated with two modes of execution that assist the guest and host to further reduce the number of interrupts and exits. The modes are NO_INTERRUPT and NO_NOTIFY (abbreviated names). When a guest turns on the NO_INTERRUPT mode, it informs the host to refrain from delivering interrupts associated with the paravirtual device until the guest turns off this mode. Virtio-net, for example, aggressively leverages this execution mode in the transmission virtqueue (Tx), because it does not typically care when the transmission finishes. Instead, virtio-net opts to recycle already-serviced Tx buffers just before posting additional buffers to be sent. Only when the virtqueue is nearly full does the guest turn off the NO_INTERRUPT mode, thereby enabling interrupts. This policy reduces the number of in-terrupts significantly.

Symmetrically, when the host turns on the NO_NOTIFY mode, it informs the guest to refrain from kicking it. The host emulation side of virtio-net routinely uses this execution mode, because TCP traffic tends to be bursty. That is, packets and frames often show up together, in bursts. Thus, when a burst of frames is sent by the guest, the host only needs the first kick. It then knows that it should start processing the ring, which is done iteratively within a loop until no more frames are left to process. While iterating, the host needs no additional kicks. Hence, it turns NO_NOTIFY on until the end of the loop and thereby minimizes virtualization overheads.

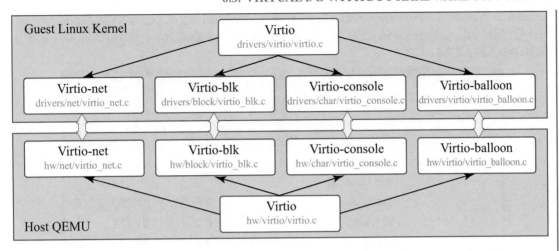

Figure 6.11: The virtio framework is used to implement paravirtual device drivers for various types of I/O, including network I/O (Ethernet), block device I/O, console character device I/O, and ballooning. Baseline virtio implements virtqueue. The individual drivers use virtqueue according to their needs. The device drivers shown are standard in Linux but do not typically ship with other OSes.

Efficiently virtualizing the network is challenging, because of the staggering throughput that modern NICs deliver, easily generating millions of packets per second, which the CPU needs to handle. For this reason, the KVM/QEMU hypervisor makes an exception for paravirtual networking. In addition to virtio-net, it also supports vhost-net [166], which breaks the traditional design of having the emulation layer entirely contained within QEMU—the userland part of the hypervisor. Vhost-net provides the same functionality as virtio-net, but instead of forwarding all the I/O events to QEMU, it handles much of the packet processing in the Linux kernel. In so doing, vhost-net nearly halves the number of context switches. Namely, virtio-net repeatedly context-switches between guest context and QEMU context via KVM (Figure 6.7), whereas vhost-net usually returns to guest context directly from KVM, without going through QEMU.

Performance: We now experimentally compare I/O paravirtualization and I/O emulation, to highlight their performance differences. Table 6.2 shows the results of running the Netperf TCP stream benchmark [108] within a Linux VM configured to access the network via e1000 or virtio-net (regular, not vhost). Netperf attempts to maximize throughput over a single TCP connection, utilizing a 16 KB message size. In our experiment, the stream traffic is outgoing from Netperf to its peer, which runs on the host system as an ordinary Linux process. (Namely, no traffic goes through a physical NIC in this experiment—only the two competing virtual NICs are involved.) The system is set to use a single core, which is saturated, spending 100% of its cycles on running the benchmark.

Table 6.2: Netperf TCP stream running in a Linux 3.13 VM on top of Linux/KVM (same version) and QEMU 2.2, equipped with e1000 or virtio-net NICs, on a Dell PowerEdge R610 host with a 2.40GHz Xeon E5620 CPU

	Metric	e1000	Virtio-net	Ratio
Guest	throughput (Mbps)	239	5,230	22x
	exits per second	33,783	1,126	1/30x
	interrupts per second	3,667	257	1/14x
TCP segments	per exit	1/9	25	225x
	per interrupt	1	118	118x
	per second	3,669	30,252	8x
	avg. size (bytes)	8,168	21,611	3x
	avg. processing time (cycles)	652,443	79,132	1/8x
Ethernet frames	per second	23,804	–	–
	avg. size (bytes)	1,259	–	–

By Table 6.2, virtio-net performs significantly better than e1000, delivering 22x higher throughput. Virtio-net's superiority is, in fact, the result of three combined improvements. Much of virtio-net's success stems from achieving its design goals of reducing the number of exits (30x less) and interrupts (14x less). Unsurprisingly, the main contributing factor responsible for the reduced overhead is the fact that, unlike e1000, virtio kicks KVM explicitly when needed, rather than implicitly triggering unintended exits due to legacy PIOs/MMIOs.

Both e1000 and virtio-net drive Ethernet NICs. Consequently, their per-frame maximum transmission unit (MTU) is defined to be 1500 bytes. Nevertheless, the NICs support TCP segmentation offload (TSO), which allows the network stack to hand to them Tx segments of up to 64KB, trusting them to break down the segments into MTU-sized frames, on their own, before transmission. The "segments per exit" metric in the table indicates that each segment requires nine exits to be processed with e1000. In contrast, with virtio-net, one exit is sufficient to transmit 25 segments, demonstrating that the NO_NOTIFY virtqueue mode, which disables kicking during bursts of activity, is highly effective. Similarly, e1000 requires one interrupt per segment, whereas virtio-net leverages this single interrupt to process 118 segments. This difference indicates that the NO_INTERRUPT virtqueue mode, which disables interrupts when possible, is likewise effective.

The quantity of exits and interrupts, however, tells only part of the story. There are two additional factors involved in determining the measured performance. First, observe that the average segment size of virtio-net is nearly 3x larger than that of e1000. The average segment size is in fact determined by the TCP/IP software stack of the guest. This stack employs a batching algorithm that aggregates messages, attempting to get more value from the NIC's TSO capability, as larger segments translate to less cycles that the core must spend on processing the message.

The profile of e1000 networking, which is much slower, discourages this sort of segment aggregation. Consequently, its segments are smaller, which translates to fewer bytes transmitted per exit/interrupt.

The other contributing factor for the performance difference is the fact that much effort went into optimizing the virtio-net code, as opposed to the e1000 code, which seems to exclusively focus on correctness rather than performance. A striking example that demonstrates this difference is the manner by which the TSO capability is virtualized. Both e1000 and virtio-net (pretend to) support this capability, exposing it to the guest OS. Virtual NICs can leverage TSO to accelerate performance in two ways: (1) if the traffic is internal to the host and does not involve sending frames over physical Ethernet (as in our benchmarking setup), then no segmentation is required, and it can be optimized away; and (2) if the traffic does flow through a physical NIC to another host system, then the TSO capability of the physical NIC can be used, if available. Wheres virtio-net employs both of these optimizations, e1000 employs neither, opting to perform segmentation in software at the QEMU emulation layer. Hence, we include the bottom part of Table 6.2, which provides statistics regrading the Ethernet frames that the e1000 emulation layer emits.

6.3.3 FRONT-ENDS AND BACK-ENDS

While this section has focused on I/O emulation and paravirtualization of the KVM/QEMU hypervisor, all production hypervisors architect their virtual I/O stacks similarly. The stack consists of two components. The first is a front-end, which encompasses a guest virtual device driver and a matching hypervisor emulation layer that understands the device's semantics and interacts with the driver at the guest. The second component is a back-end, which is used by the front-end to implement the functionality of the virtual device using the physical resources of the host system. Among the two components, the front-end serves as the interface that is exposed to the virtual machine, and the back-end serves as the underlying implementation. Importantly, the two components are architected to be independent of each other. Thus, production hypervisors allow users to compose their virtual machines from different front-ends and back-ends.[4]

Figure 6.12 shows the emulated network front-ends and back-ends made available by QEMU [148]. We have already mentioned e1000 and virtio-net. The rtl8139 corresponds to an emulation of the Realtek RTL8139C(L)+ 10/100M Fast Ethernet Controller [150]. In terms of back-ends, TAP is a virtual Ethernet network device, implemented in the kernel, that can forward Ethernet frames to and from the process that connects to it. The Ethernet frames can also be forwarded using a regular user-level socket, or over a Virtual Distributed Ethernet (VDE) [174]. The MACVTAP back-end accelerates the performance of vhost-net by eliminating an ab-

---

[4]The terms front-end and back-end, when applied to I/O virtualization, are overloaded. They are used differently and inconsistently by different people in different contexts. The definitions presented here are consistent with the QEMU documentation [148] and with the authoritative survey of Waldspurger and Rosenblum on I/O virtualization [178].

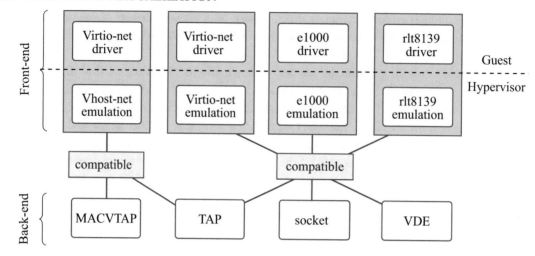

Figure 6.12: The QEMU network front-ends and back-ends are composable: virtual machines can be configured to use a certain front-end and any of its compatible back-ends.

straction layer within the Linux kernel at the cost of losing routing flexibility (instead of going through a software bridge, the vhost-net traffic flows directly to a NIC) [167].

Other device types offer different front-ends and back-ends. For example, when considering storage, a VM may be configured to use, say, a SCSI, IDE, or paravirtual block device front-ends, and a local or remote file back-end.

> In summary, hypervisors employ I/O emulation by redirecting guest MMIOs and PIOs to read/write-protected memory. Emulation exposes virtual I/O devices that are implemented in software but support hardware interfaces. Paravirtualization improves upon emulation by favoring interfaces that minimize virtualization overheads. The downside of paravirtualization is that it requires hypervisor developers to implement per-guest OS drivers, and it hinders portability by requiring users to install hypervisor-specific software.

6.4   VIRTUAL I/O WITH HARDWARE SUPPORT

In the beginning of §6.3, when we started the discussion about how virtual I/O is implemented, we ruled out the possibility to allow VMs to directly access physical devices, because they are not aware of the fact that they must share, which would undoubtedly result in undesirable behavior and possibly unrecoverable damage. This difficulty motivated device emulation and paravirtualization, which incur virtualization overheads but provide safe virtual I/O and the benefits of I/O

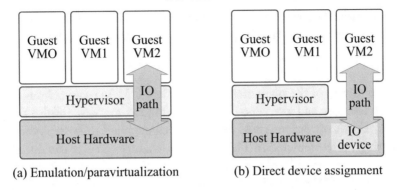

(a) Emulation/paravirtualization   (b) Direct device assignment

Figure 6.13: Direct device assignment exclusively dedicates an I/O device to a specific VM, improving performance by reducing virtualization overheads at the cost of forgoing I/O interposition.

interposition outlined in §6.1. But if we are willing to forego I/O interposition and its many advantages, there is in fact another, more performant alternative that reduces the overheads of virtualization. Given a physical device d, the hypervisor may decide to assign the right to access d exclusively to some specific virtual machine v, such that no other VM, and not even the hypervisor, are allowed access; the hypervisor can do so if it has an extra physical device that it does not strictly need. This approach, denoted **direct device assignment**, is illustrated in Figure 6.13. Seemingly, in such a setup, v can safely access d directly, since nobody else is using it. The upshot would be better performance for v, because the overhead of virtualization would be significantly reduced.

Such naive device assignment may work in certain contexts, but it suffers from two serious problems. The immediate one is lack of scalability. Extra physical I/O devices, if they exist, are inherently limited in quantity. The number of virtual machines that a modern server can support is much larger than the number of physical I/O devices that it can house. The second drawback is less obvious but more severe. If v controls a device, then v can program the device to perform DMA operations directed at *any* physical memory location. In other words, by assigning d to v, we allow v to (indirectly) access the entire physical memory, including areas belonging to other VMs or the hypervisor. Therefore, by utilizing naive direct device assignment, we essentially eliminate isolation between VMs. We thus have to trust v not to abuse its power, maliciously or mistakenly, which is overwhelmingly unacceptable.

Hardware support for I/O virtualization allows us to resolve the aforementioned two problems, and this section is dedicated to overviewing how. Specifically, the first part of the section describes the hardware mechanisms that are able make direct device assignment both secure and scalable. Security is obtained with the help of the **I/O Memory Management Unit** (IOMMU), and scalability is obtained via **Single-Root I/O Virtualization** (SRIOV). The combination of SRIOV and IOMMU makes device assignment a viable performant approach, leaving one last major source of virtualization overhead: interrupts. The second part of this section therefore describes the hardware mechanisms that help hypervisors mitigate this overhead.

6.4.1   IOMMU

Recall that device drivers of the operating system initiate DMAs to asynchronously move data from devices into memory and vice versa, without having to otherwise involve the CPU (§6.3). Not so long ago, DMA operations could have only been applied to host-physical memory addresses. In the ring buffer depicted in Figure 6.4, for example, this constraint means that the driver must use physical target buffers when populating the ring's descriptors. The device then uses these addresses for DMA in their raw form—they do not undergo address translation.

As noted, this DMA policy essentially means that virtual machines with assigned devices can indirectly read/write any memory location, effectively making MMU-enforced isolation meaningless. A second, less apparent problem is that such VMs can also indirectly trigger any interrupt vector they wish (explained below). The third problem has more of a technical nature. VMs do not know the (real) physical location of their DMA buffers, since they use guest-physical rather than host-physical addresses (Figure 5.1). Consequently, to be able to use DMAs of directly assigned devices, VMs must somehow become aware of the fact that their "physical" memory is not actually physical, and they need the hypervisor to expose to them the underlying host-physical addresses.

To address all these problems, all major chip vendors introduced I/O memory management units (IOMMUs), notably in Intel's Virtualization Technology for Directed I/O (VT-d) [106], AMD's I/O Virtualization Technology (AMD-Vi) [11], ARM's System Memory Management Unit Architecture Specification (IOMMU is called SMMU) [23], IBM POWER [93, 94], and Sun (later Oracle) SPARC [164]. While the IOMMU implementations differ, they have a similar functionality. Recall that this chapter focuses on VT-d.

The IOMMU consists of two main components: a DMA remapping engine (DMAR) and an interrupt remapping engine (IR). DMAR allows DMAs to be carried out with I/O virtual addresses (IOVAs), which the IOMMU translates into physical addresses according to page tables that are set by the hypervisor. Likewise, IR translates interrupt vectors fired by devices based on an interrupt translation table configured by the hypervisor. IOVAs live in I/O address spaces that can overlap, be the same, or be completely different than host or guest address spaces.

DMA Remapping: To simplify the explanation of DMAR, let us first consider bare metal setups, where there are only regular processes and no virtual machines. In such setups, the role the IOMMU plays for I/O devices is similar to the role the regular MMU plays for processes, as illustrated in Figure 6.14a. Processes access the memory using virtual addresses, which are translated into physical addresses by the MMU. Analogously, I/O devices access the memory via DMAs associated with IOVAs, which are translated into physical addresses by the IOMMU. The OS controls how IOVAs get translated similarly to the manner by which it controls how regular virtual addresses get translated. Specifically, given a target memory buffer of a DMA, the OS associates

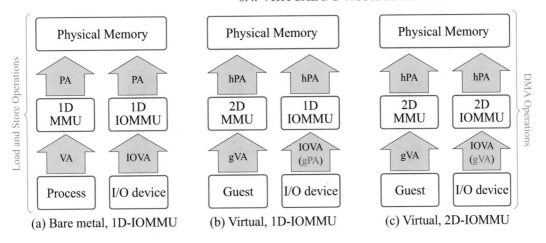

(a) Bare metal, 1D-IOMMU   (b) Virtual, 1D-IOMMU   (c) Virtual, 2D-IOMMU

Figure 6.14: In non-virtual setups, the IOMMU is for devices what the MMU is for processes (a). When virtual machines are involved, the usage model of the IOMMU (for facilitating direct device assignment) depends on whether the IOMMU supports 1D or 2D page walks. With a 1D IOMMU, the hypervisor does not expose an IOMMU to the guest, causing it to program DMAs with guest-physical addresses (b). With a 2D IOMMU, the guest has its own IOMMU, to be programmed as it pleases (c).

the physical address (PA) of that buffer with some IOVA. Then, the OS maps the IOVA to the PA by inserting the IOVA⇒PA translation into IOMMU data structures.

Figure 6.15 depicts these structures for Intel x86-64 [106]. PCIe dictates that each DMA will be associated with the 16-bit bus-device-function (BDF) number that uniquely identifies the corresponding I/O device in the PCIe hierarchy (as described in §6.2.3). The DMA propagates upstream, through the hierarchy (Figure 6.5), until it reaches the root complex (RC) where the IOMMU resides. The IOMMU uses the 8-bit bus identifier to index the root table in order to retrieve the physical address of the context table. It then indexes the context table using the 8-bit concatenation of the device and function identifiers. The result is the physical location of the root of the page table hierarchy that houses all the IOVA⇒PA translations of this particular I/O device (PCIe function).

The IOMMU walks the page table similarly to the MMU, checking for translation validity and access permissions at every level. If the translation is invalid, or the access permissions mismatch, the IOMMU fails the translation. Similarly to how the MMU utilizes its TLB, the IOMMU caches translations using its IOTLB. However, different than the MMU, the IOMMU typically does not handle page faults gracefully. I/O devices usually expect their DMA target buffers to be present and available, and they do not know how to recover otherwise. (Recall that before IOMMUs, devices worked with physical addresses; even with IOMMUs, devices still "believe" that the DMA addresses they are given are physical.) For this reason, when the hypervisor

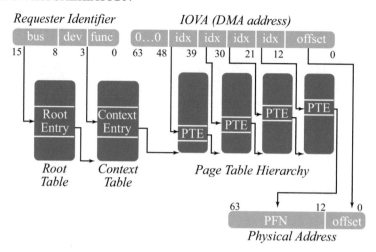

Figure 6.15: IOVA translation with the Intel IOMMU.

directly assigns a device to a virtual machine, it typically pins the entire address space of the VM to the physical memory, to avoid DMA page faults. In so doing, the hypervisor loses its ability to apply the canonical memory optimizations to the address space of the VM, as discussed in §6.1.

Let us now consider a virtual setup with virtual machines. As of this writing, overwhelmingly, modern Intel servers support a two-dimensional (2D) MMU (see §5.1) but only one-dimensional (1D) IOMMU. Recall that a 2D-MMU allows the guest and hypervisor to maintain their own page tables, such that: (1) the guest OS page tables map guest-virtual addresses (gVAs) to guest-physical addresses (gPAs); (2) the hypervisor page tables map guest-physical addresses to host-physical addresses (hPAs); and (3) the processor makes use of the two page table levels for translating gVAs all the way to hPAs, as depicted in the left of Figure 6.14b. When directly assigning a device to a guest, the hypervisor copes with having an IOMMU that is only 1D by *not* exposing an IOMMU to the guest. The guest then believes that it must program DMA operations with physical addresses, which are in fact gPAs. To make this work, when mapping the address space of the guest in the IOMMU, the hypervisor uses IOVAs identical to the guest's gPAs, such that the guest memory addresses and its (identical) I/O addresses point to the same host-physical location, as depicted in the right of Figure 6.14b.

Not exposing an IOMMU to the guest is suboptimal in several respects. Similarly to the MMU, it makes sense to virtualize the IOMMU, such that guest and host would directly control their own page tables, and the hardware would conduct a 2D page-walk across the two levels, as depicted in Figure 6.14c. 2D IOMMUs allow guests to protect themselves against errant or malicious I/O devices—similarly to bare metal OSes—by mapping and unmapping the target buffer of each DMA right before the DMA is programmed and right after it completes. This policy is called strict IOMMU protection [125, 142], and it is recommended by hardware vendors and operating system developers [106, 109, 131, 183]. Another benefit of 2D IOMMUs is that they help guests to use legacy devices that, for example, do not support memory addresses wider than 32 bit. Such a constraint is potentially problematic, but it can be easily resolved by programming the IOMMU to map the relevant 32 bit-addresses to higher memory locations. A third bene-fit of 2D IOMMUs is that they allow guests to directly assign devices to their own processes, allowing for user-level I/O, which is lately becoming increasingly popular [98]. 2D IOMMUs have been thus far implemented with the help of software emulation [12]. But they are already defined by the x86-64 I/O virtualization specifications of Intel [106] and AMD [11]. As of this writing, it seems that 2D IOMMU support can be enabled in certain Broadwell and Skylake Intel processors [122, 186].

Interrupt Remapping: Recall that PCIe defines (MSI/MSI-X) interrupts similarly to DMA memory writes directed at some dedicated address range, which the RC identifies as the "inter-rupts space" (see §6.2.3). In x86, this range is 0xFEEx_xxxx (where x can be any hexadecimal digit). Each interrupt request message is self-describing: it encodes all the information required for the RC to handle it.

    To understand why interrupt remapping is necessitated, let us first consider how interrupt delivery of an assigned device d, which is given to a virtual machine v, works without IR (illus-trated in Figure 6.16). Setting the interrupt vector ("data") and target LAPIC ("address") that are associated with d requires the operating system to MMIO-write these data and address into the MSI fields of d's configuration space. But since v is a virtual machine, its configuration space is emulated: it is not allowed to access the physical configuration space, which is (or can be per-ceived as) a singular device that affects the entire system. Assume that the hypervisor decided that d's (physical) interrupt vector is 50, and that, later, v configured this vector to be 30 via its emu-lated configuration space (Figure 6.16a). Therefore, the configuration space emulation layer of the hypervisor records internally that v expects to get a vector of 30 from d, rather than 50. Hence-forth, when d fires an interrupt (vector 50), the corresponding MSI message passes through the IOMMU unchanged, because there is no IR (Figure 6.16b). It then reaches the target LAPIC, thereby triggering an exit that invokes the hypervisor, because the latter configured v's VMCS to exit upon interrupts (Figure 6.16c). When awakened, the hypervisor delivers the interrupt to v using vector 30, as indicated by v's emulated configuration space (Figure 6.16d).

    The above interrupt delivery procedure is seemingly safe. But in fact it can be exploited by v to arbitrarily change the physical interrupt vector delivered to the hypervisor (Figure 6.16c), i.e., to values different than 50 [185]. Because d is directly assigned to v, the latter can program the former to DMA-write any value (any vector) into the interrupt space, by using 0xFEEx_xxxx as a DMA target address. Crucially, without IR, the IOMMU cannot distinguish between a legitimate, genuine MSI interrupted fired by d and a rogue DMA that just pretends to be an interrupt.

    Figure 6.17 demonstrates how the IR capability of the IOMMU eliminates the vul-nerability. First, the type and format of the MSI registers in the configuration space change

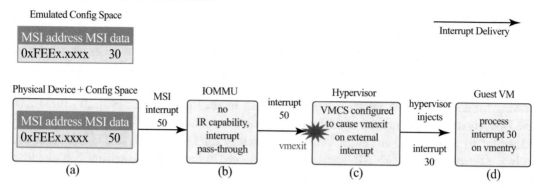

Figure 6.16: MSI interrupt delivery without interrupt remapping support.

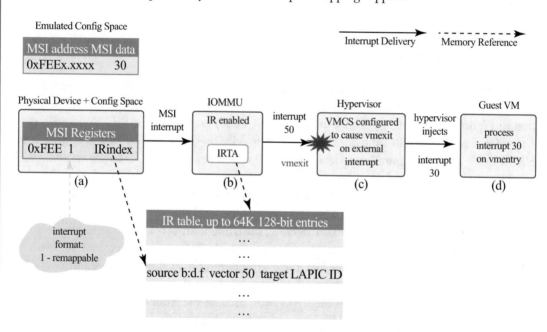

Figure 6.17: MSI interrupt delivery with interrupt remapping support. (IRindex is denoted "interrupt_index" in the VT-d specification.)

(Figure 6.17a, to be compared with Figure 6.16a). Instead of storing the physical interrupt vector and target LAPIC, MSI registers now store an IRindex, which is an index to the IR table, pointed to by the IR Table Address (IRTA) register at the IOMMU. As before, the content of MSR registers in the physical configuration space is exclusively set by the hypervisor. When d fires an interrupt directed at 0xFEEx_xxxx, the IOMMU deduces the IRindex form this "address" and "data" (Figure 6.17b). The IOMMU then determines the target LAPIC and physical interrupt vector based on the associated IR table entry. The rest of the interrupt delivery flow is identical to the flow depicted in Figure 6.16.

In addition to the target LAPIC and interrupt vector, the IR table entry contains d's BDF for anti-spoofing, to indicate that d is indeed allowed to raise the interrupt associated with this IRindex. Only after verifying the authenticity and legitimacy of the interrupt, does the IOMMU deliver it to the hypervisor, as before. Since the target LAPIC and interrupt vector are determined based on the IR table—which is set by the hypervisor—they cannot be forged. Although v can still program DMAs that look like interrupts, they will only be delivered to v as the hypervisor intended, and they will not trick the IOMMU into issuing the wrong physical interrupt vector.

6.4.2   SRIOV

With IOMMUs making direct device assignment safe, we go on to discuss how hardware support (in the PCIe and device level) additionally makes it scalable. Since (1) physical servers can house only a small number of physical devices as compared to virtual machines, and because (2) it is probably economically unreasonable to purchase a physical device for each VM, the PCI-SIG standardized the SRIOV specification, which extends PCIe to support devices that can "self-virtualize". Namely, an SRIOV-capable I/O device can present multiple instances of itself to software. Each instance can then be assignment to a different VM, to be used directly (and exclusively) by that VM, without any software intermediary. Traditionally, it is the role of the operating system to multiplex the I/O devices. Conversely, an SRIOV device knows how to multiplex itself at the hardware level.

An SRIOV device is defined to have at least one Physical Function (PF) and multiple Virtual Functions (VFs), which serve as the aforementioned device instances, as illustrated in Figure 6.18. A PF is a standard PCIe function (defined in §6.2.3). It has a standard configuration space, and the host software manages it as it would any other PCIe function. In addition to supporting the standard operations, the PF also allows the host to allocate, deallocate, and configure VFs. A VF is a lightweight PCIe function that implements only a subset of the components of a standard PCIe function. For example, it does not have its own power management capabilities and instead shares such capabilities with its PF, and it cannot (de)allocate other VFs. Accordingly, it has a limited configuration space that presents limited capabilities. When a VF is assigned to a virtual machine, the former provides the latter the ability to do direct I/O—the VM can safely initiate DMAs, such that the hypervisor remains uninvolved in the I/O path. The theoretical bound on the number of VFs that can be exposed by a single physical device is 64 K. Current Intel and Mellanox implementations of SRIOV NICs enable up to 128 and 512 VFs per device, respectively [105, 128].

Using the lspci shell command, Figure 6.19 shows some of the content of the configuration space of an X540 Intel NIC, which supports SRIOV. As can be seen, the output indeed includes the SRIOV standard PCIe extended capability. The output further specifies the maximal number of VFs that this NIC supports (64), and the current number of allocated VFs (0). Another

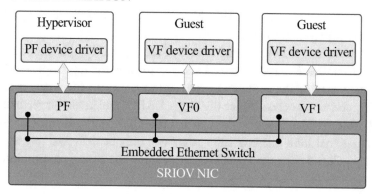

Figure 6.18: SRIOV-capable NIC in a virtualized environment.

```
06:00.0 Ethernet controller: Intel Ethernet Controller 10-Gigabit X540-AT2
        Subsystem: Intel Corporation Ethernet 10G 2P X540-t Adapter
        Flags: bus master, fast devsel, latency 0
        Memory at 91c00000 (64-bit, prefetchable) [size=2M]
        Memory at 91e04000 (64-bit, prefetchable) [size=16K]
        Expansion ROM at 91e80000 [disabled] [size=512K]
        Capabilities: [70] MSI-X: Enable+ Count=64 Masked-
        Capabilities: [a0] Express Endpoint, MSI 00
        Capabilities: [150] Alternative Routing-ID Interpretation (ARI)
        Capabilities: [160] Single Root I/O Virtualization (SR-IOV)
                Total VFs: 64, Number of VFs: 0
                VF offset: 128, stride: 2, Device ID: 1515
                Supported Page Size: 00000553, System Page Size: 00000001
                Region 0: Memory at 92300000 (64-bit, non-prefetchable)
                Region 3: Memory at 92400000 (64-bit, non-prefetchable)
        Capabilities: [1d0] Access Control Services
```

Figure 6.19: Partial lspci output of the Intel X540 NIC shows the SRIOV and ARI capabilities.

relevant capability that the NIC supports is Advanced Routing-ID Interpretation (ARI) [139]. Recall that PCIe enumerates bridges and device endpoints with a 16-bit bus:device.function (BDF) identifier, such that bus, device, and function consist of 8, 5, and 3 bits. The ARI capability lets SRIOV devices support more functions at endpoints than just the eight that the BDF enumeration permits (3 bits). In particular, ARI allows manufacturers of SRIOV-capable devices to omit the endpoint's 5-bit device identifier for this purpose. The 16-bit identifier of such endpoints is bus.function, such that both bus, and function consist of 8 bits, thereby supporting up to 256 functions per endpoint.

In §6.2.3, we described the PCIe hierarchy and its BDF enumeration. SRIOV changes the manner by which this enumeration is performed. Consider the enumeration depicted in

Figure 6.5, and assume that we now replace the device that houses PCIe function 5:0.0 with an SRIOV device d that supports one PF and up to 2048 VFs (similarly to the the Altera Arria 10 [10]). Further assume that d uses ARI (which requires that all its ancestors in the PCIe tree also support ARI). Figure 6.20 highlights the differences in the BDF tree due to the change we applied, as compared to Figure 6.5. In the example shown, 600 of the 2048 VFs of d are allocated; the remaining VFs can be allocated later. We can see that all the rules that govern BDF enumeration still apply. The difference is that the enumeration construction process (conducted by the BIOS/UEFI, and perhaps also later by the OS while booting) queries the configuration space of devices and, if they are SRIOV-capable, creates a "hole" in the hierarchy that is big enough to accommodate all their possible future VF creations. A PCIe bus can connect to 256 endpoints, so d requires reserving nine buses: eight to account for the 2048 potential VFs and another for the single PF (PFs and VFs can reside on the same bus; the nine's bus is needed because eight are not enough). Consequently, Buses 5–13 are reserved for d, which means function 6:0.0 from Figure 6.5 becomes 14:0.0 in Figure 6.20.[5]

Figure 6.21 compares the network performance of SRIOV device assignment to that of e1000 emulation (§6.3.1) and virtio-net paravirtualization (§6.3.2), by running Netperf stream inside a virtual machine, which uses the three I/O models to communicate with a remote machine. The underlying physical NIC is SRIOV-capable, and the traffic of all three I/O models eventually flows through it. However, the e1000 and virtio-net front-ends are implemented/mediated by the hypervisor, which uses the SRIOV-capable NIC as a back-end. In contrast, the VM in the SRIOV experiment uses a VF, thereby keeping the hypervisor uninvolved when it interacts with its NIC instance through the corresponding ring buffers.

The results in Figure 6.21 indicate that using SRIOV is indeed advantageous in terms of performance, improving the throughput of the VM by 1.31–1.78x relative to paravirtualization, depending on the message size. The results further show that, as expected, the throughput is inversely proportional to the number of exits—fewer exits translate to reduced virtualization overheads, allowing the VM to spend additional CPU cycles to push more data into the network.

6.4.3   EXITLESS INTERRUPTS

With SRIOV and the IOMMU, the hypervisor can safely and scalably assign devices directly to guests. But direct device assignment does not eliminate all the major sources of virtualization overhead. Whereas it eliminates most of the exits that occur when guests "talk" to their assigned devices (via MMIOs, for example), it does not address the overheads generated when assigned devices "talk back" by triggering interrupts to notify guests regarding the completion of their I/O requests. Intel VT-x (rather than VT-d) provides the hypervisor with the basic hardware support needed to inject virtual interrupts into guests. Next, we outline this support, asses its overheads,

---

[5]In Linux, users can easily create and destroy VFs by writing the desired number of VFs to the sriov_numvfs file that is associated with the device, which is found under the /sys file system. For example, "echo 10 > /sys/class/net/eth1/device/sriov_numvfs".

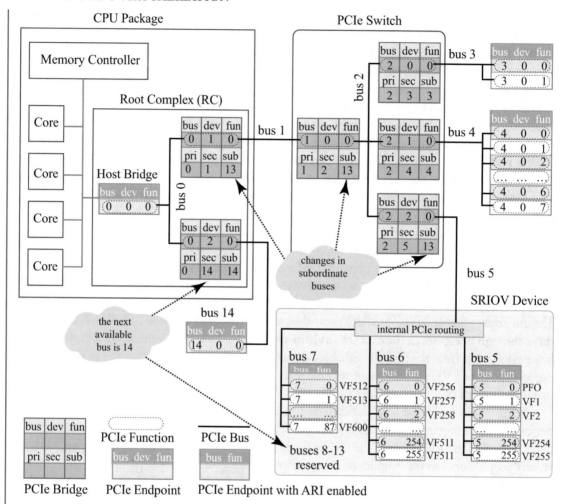

Figure 6.20: The process of BDF enumeration of PCIe bridges and endpoints changes in the presence of SRIOV-capable devices, creating "holes" in the hierarchy to accommodate for the creation of VFs (compare with Figure 6.5).

show that they are significant, and describe software—and later hardware—means to alleviate this problem.

Basic VT-x Interrupt Support: Similarly to other CPU registers, the VMCS stores the value of the guest's Interrupt Descriptor Table Register (IDTR, defined in 6.2.1), such that it is loaded upon entering guest mode and saved on exit, when the hypervisor's IDTR is loaded instead.

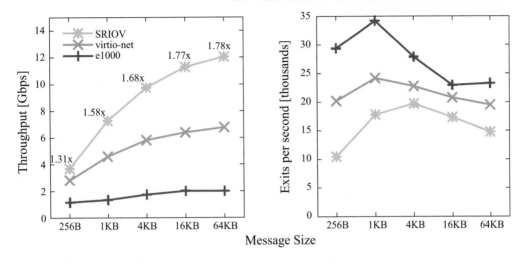

Figure 6.21: Netperf TCP stream running in a single-VCPU Linux 4.4.2 VM served by Linux/KVM (same version) and QEMU 2.5 on a Dell PowerEdge R420 host equipped with a 1.9 GHz Xeon E5-2420 CPU and a 40 Gbps SRIOV-capable Mellanox ConnectX-3 NIC. Networking of the VM is done through e1000, virtio-net (with vhost-net and MACVTAP), and a VF of the ConnectX-3. The destination is a remote R420, which does not employ virtualization. The numbers show the ratio of virtio-net to SRIOV.

Roughly speaking, control over all interrupts implies control over the CPU.[6] The hypervisor needs to maintain control, rather than surrender it to the untrusted guest, so it sets a VMCS control bit (denoted "external-interrupt exiting"), which configures the core to exit whenever an external interrupt fires. The chain of events that transpire when such an event occurs is depicted in Figure 6.22.

Assume that the guest is computing (i), but then an I/O device raises an MSI/MSI-X interrupt (ii), causing an exit that transfers control to the hypervisor (iii). Using the "VM-exit information" fields in the VMCS, the hypervisor figures out that the cause for the exit was an external interrupt. It therefore handles the interrupt and acknowledges the completion of the interrupt handling activity through the LAPIC End of Interrupt (EOI) register, as required by the x86 specification (iv). In our example, the physical interrupt is associated with the guest, so the hypervisor injects an appropriate virtual interrupt (possibly with a different vector) to the guest (v). Technically, the hypervisor injects by writing to the "interruption-information" field (that resides in the "VM-entry control" area of the VMCS) the type of the event (external interrupt) and its vector. The actual interrupt delivery takes place within the guest during vmentry, when the guest's state—including its IDTR—is already loaded to the CPU, thereby allowing the guest to handle

---

[6]The hypervisor can regain control by using the preemption time feature of x86 virtualization, which triggers an unconditional exit after a configurable period of time elapses.

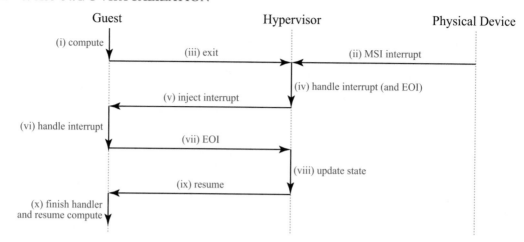

Figure 6.22: Chain of events when a physical device triggers an interrupt, without hardware support for direct interrupt delivery.

the interrupt (vi). Shortly after, the guest acknowledges interrupt completion, which, again, results in an exit, because the hypervisor emulates the LAPIC for the guest (vii). The hypervisor updates its internal state accordingly (viii) and resumes guest execution (ix). The guest can now complete the handler routine and return to normal execution (x).

To complete the description of the basic VT-x interrupt support, we note in passing that it is possible for the guest to be in a state whereby it is not ready to receive an interrupt, notably because its interrupt flag is off. In this case, the hypervisor sets the "interrupt-window exiting" VMCS bit, which will cause the core to trigger an exit when the guest is ready to receive interrupts yet again, at which point the hypervisor will be able to inject as usual.

Assigned EOI Register: VT-x interrupt support, as described above, requires at least two exits per interrupt. Thus, when devices generate many interrupts per second, the overhead of virtualization can be substantial. The question is whether we can alleviate this overhead. Let us first focus on eliminating the exits associated with EOI. The operating system uses the LAPIC to control all interrupt activity (§6.2.1). To this end, the LAPIC employs multiple registers used to configure, block, deliver, and (notably in this context) signal EOI. If we assume that an interrupt of a physical device can somehow be safely delivered directly to the guest without hypervisor involvement (discussed below), then the EOI register should correspondingly also be assigned to the guest. EOI assignment, however, is not possible with older LAPIC generations, because they expose all the LAPIC registers in a tightly packed predefined memory area that is accessed through regular load and store instructions. Therefore, with an emulated LAPIC, providing the guest with write permissions to one register implies that the guest can write to all registers that reside in the same page, as memory protection is conducted in page granularity.

Thankfully, the current LAPIC interface, x2APIC, exposes its registers using model specific registers (MSRs), which are accessed through "read MSR" and "write MSR" instructions. The CPU exits on LAPIC accesses according to an MSR bitmap controlled by the hypervisor. The bitmap specifies the "sensitive" MSRs that cannot be accessed directly by the guest and thus trigger exits. In contrast to other LAPIC registers, with appropriate safety measures [13], EOI can be securely assigned to the guest.

Assigned Interrupts: Assume a guest is directly assigned with a device. By utilizing a software-based technique called exitless interrupts (ELI) [80], it is possible to additionally securely assign the device's interrupts to the guest—without modifying the guest or resorting to paravirtualization—by employing the architectural support described thus far. An assigned exitless interrupt does not trigger an exit. It is delivered directly to the guest without host involvement. Consequently, ELI eliminates most of the remaining virtualization overhead.

ELI is structured based on the assumption that in high-performance, SRIOV-based device assignment deployments nearly all physical interrupts arriving to a given core are targeted at the guest that runs on that core. By turning off the aforementioned VMCS "external-interrupt exiting" control bit, ELI delivers all physical interrupts on the core to the guest running on it. At the same time, ELI forces the guest to reroute to the hypervisor all the interrupts that are not assigned to it, as follows. While the guest initializes and maintains its own IDT, ELI runs the guest with a different IDT—called shadow IDT—which is prepared by the hypervisor. Just like shadow page tables can be used to virtualize the guest MMU without any cooperation from the guest (§4.2.3), IDT shadowing can be used to virtualize interrupt delivery. Using the trap and emulate technique, the hypervisor monitors the updates that the guest applies to its emulated, write-protected IDT, and it reacts accordingly so as to provide the desired effect.

The ELI mechanism is depicted in Figure 6.23. By shadowing the guest IDT, the hypervisor has explicit control over which handlers are invoked upon interrupts. It thus configures the shadow IDT to (1) deliver assigned interrupts directly to the guest's interrupt handler, and to (2) force an exit for non-assigned interrupts by marking the corresponding IDT entries as non-present.

Performance: Figure 6.24 (left) shows the benefit of employing the ELI technique—which encompasses EOI assignment—utilizing Netperf TCP stream (with 256 B message size), Apache [18, 69] (HTTP server driven by ApacheBench [17]), and Memcached [70] (key-value storage server driven by Memslap [9]) to communicate with a remote client machine via an SRIOV-capable Emulex OneConnect 10Gbps NIC. Each of the benchmarks is executed on bare-metal and under two virtualized setups: the SRIOV device assignment baseline, and SRIOV supplemented with ELI. We can see that baseline SRIOV performance is considerably below bare-metal performance: Netperf's VM throughput is at 60% of bare-metal throughput, Apache is at 65%, and Memcached is at 60%. It is therefore evident that using ELI gives a significant throughput increase over baseline SRIOV: 63%, 49%, and 66% for Netperf, Apache, and Memcached, respectively. Importantly, with ELI, Netperf achieves 98% of bare-metal throughput,

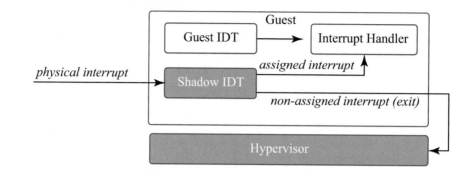

Figure 6.23: ELI interrupt delivery flow.

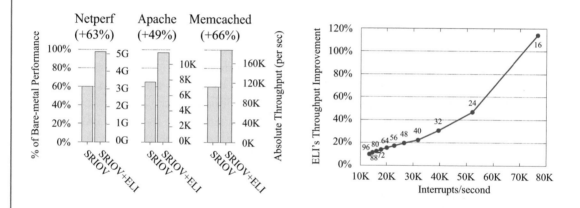

Figure 6.24: Left: throughput of three SRIOV benchmarks, with/without ELI, shown as percentage of bare-metal performance (left y axis) and in absolute terms (right y axis); numbers under the titles show ELI's relative improvement over SRIOV. Right: Netperf's relative performance improvement gained with ELI as a function of interrupts per second; labels along the curve specify the NIC's interrupt coalescing interval in microseconds. Data taken from the ELI paper [80], utilizing a single-VCPU Linux 2.6.35 VM served by Linux/KVM (same version) and QEMU 0.14 on an IBM System x3550 M2 server equipped with a 2.93 GHz Xeon X5570 CPU and an Emulex OneConnect 10 Gbps NIC.

Apache 97%, and Memcached nearly 100%. Namely, ELI all but eliminates virtualization overheads.

We now proceed to investigate the dependence of ELI's improvement on the amount of interrupt coalescing done by hardware (NIC), which immediately translates to the amount of generated interrupts. (We note that software interrupt mitigation techniques, employed by the Linux kernel, are also in affect throughput the experiments.) The Emulex OneConnect NIC imposes a configurable cap on coalescing, allowing its users to set a time duration T, such that the

NIC will not fire more than one interrupt per $T\mu$s (longer T implies less interrupts). The available coalescing cap values are: 16 μs, 24 μs, 32 μs, 40 μs, ..., 96 μs. Figure 6.24 (right) plots the results of the associated experiments, such that the labels along the curve denote the corresponding value of T. Unsurprisingly, higher interrupt rates imply higher savings due to ELI. The smallest interrupt rate that the NIC generates for this workload is 13 K interrupts/sec (with T=96 μs), but even with this maximal coalescing, ELI still provides a 10% performance improvement over baseline SRIOV. ELI additionally achieves nearly 100% of bare-metal throughput in all of the experiments shown in Figure 6.24 (right), indicating that, when it is used, coalescing has a lesser effect on throughput. Granularity of coalescing can therefore be made finer as needed, so as to refrain from the increased latency that coarse coalescing induces (not shown).

6.4.4 POSTED INTERRUPTS

We have seen that the overhead of interrupts and EOIs can be substantial in virtualized environments. ELI combats this problem by using software-based interrupt and EOI assignment. While effective, ELI has two notable drawbacks. First, it increases hypervisor complexity, not just because of IDT shadowing, but also because of the security measures that the hypervisor must employ [80] to prevent malicious or buggy guests from exploiting the technique. For example, the guest might decide never to acknowledge completion of interrupts, which affects the hypervisor, since guest and hypervisor share the physical LAPIC EOI register. The second drawback is that ELI is inherently an all-or-nothing approach: either all interrupts go initially to the guest, which might adversely affect, for example, the interrupt latency experienced by the hypervisor, or ELI cannot be used.

Unlike all other virtualization techniques discussed in this chapter, ELI, which is the product of a fairly recent academic exercise [13, 80], has never been widely deployed—hardware support that deems it unnecessary was quick to emerge.[7] Ideally, we would like such support to allow the hypervisor to easily assign specific interrupts of specific devices to specific guests, and to have all other interrupts delivered directly to the host without requiring any sophisticated software hacks. Posted interrupts is a recently introduced mechanism that provides such support. It divides into two components: (1) CPU posted interrupts, which correspond to IPIs, namely, interrupts that are directly injected by the hypervisor that runs on one core to a guest that runs on a different core without involving the hypervisor on the latter core, and (2) IOMMU posted interrupts (denoted "VT-d posted interrupts" by Intel), which correspond to interrupts that are delivered directly from I/O devices to guest VMs. Both components rely on Intel's APIC virtualization (APICv), which provides a "virtual APIC" for the guest, whose semantics are preserved by the underlying hardware, rather than by the hypervisor via LAPIC emulation. Using the vir-

---

[7]We nevertheless include ELI in this book to motivate such hardware support, as we do not yet have performance numbers associated with the latter. To our knowledge, Intel processors that support functionality similar to that of ELI ("VT-d posted interrupts," to be discussed shortly) have started to ship only recently, in 2016. AMD has likewise recently added similar functionality, called Advanced Virtual Interrupt Controller (AVIC).

tual APIC, the hypervisor can configure guests to receive and acknowledge the completion of interrupts without involving the hypervisor on the guest's core [107].

Virtual APIC: The hypervisor configures the VMCS of the guest to point to a 4 KB memory area, denoted as the "virtual APIC page," which the processor uses in order to virtualize access to APIC registers and track their state, and to manage virtual interrupts. The page houses all the registers of the guest's virtual APIC. Recall that a physical APIC has an IRR, ISR, EOI, and ICR registers (their meaning and connections are described in §6.2.1). Thus, correspondingly, the processor maintains a virtual version of these registers in the virtual APIC page (vIRR, vISR, vEOI, vICR), at the same offsets as their physical counterparts. When a virtual register is updated, the hardware emulates the side-effects that would have occurred if a physical APIC register was updated similarly; this behavior is called "APIC-write emulation." For example, when the guest acknowledges interrupt completion in its vEOI, the hardware removes the associated interrupt bit from the vISR and then, e.g., delivers to the guest the highest-priority pending interrupt, updating vIRR and vISR accordingly. Only after the APIC-write emulation is performed does the hardware trigger exits, if necessary. In our example, if the hypervisor configured the VMCS to trigger exits upon EOI updates, the vISR is updated and only then the exit takes place.

APICv allows the hypervisor to associate a guest with a fully functioning hardware-supported virtual APIC, which is of course different than the physical APIC. Below, we outline how to trigger interrupts directed at a specific guest virtual APIC, while keeping the hypervisor at the target core uninvolved. The subsequent (virtual) EOI operation would likewise only affect this particular virtual APIC and its guest. The interrupt handling operations of the virtual APIC are completely decoupled from the hypervisor's physical interrupt activity. Generally speaking, APICv provides an interface for compute entities (a different core, an SRIOV device) to turn on bits in designated virtual APIC spaces and have the system behave as if an interrupt was generated for the associated guest, and only for it.

CPU Posted Interrupts: Let us describe the delivery flow of CPU posted interrupts through an example, which also highlights how virtual APICs work. The example is depicted in Figure 6.25 and explained in detail in the text that follows. Let R denote the value of an interrupt vector, which the hypervisor on Core 1 would like to deliver *directly* to a guest that currently executes on Core 2, without causing an exit on Core 2. The hypervisor must initially update the guest's posted interrupt descriptor ("PI descriptor"), which is pointed to by the guest's VMCS (Figure 6.25a). Ordinarily, for correctness, hypervisor software must not modify any of the data structures pointed to by the VMCS while the associated guest is running. But this requirement does not apply to the PI descriptor, so long as it is modified using an atomic read-modify-write instruction. The format of the PI descriptor is specified in Table 6.3 [106] and explained next. The hypervisor uses the PI descriptor's first field—the posted-interrupt requests (PIR) bitmap—to set the bit that identifies the interrupt vector R that the hypervisor wishes to directly deliver to the guest. In our example, R=30.

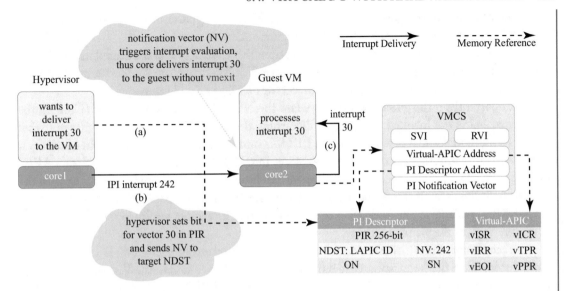

Figure 6.25: CPU posted interrupt illustration.

Table 6.3: The posted interrupts (PI) descriptor consists of 512 bits (64 bytes)

Bits	Name	Description
255:0	PIR	Post-Interrupt Requests, one bit per requested vector
256	ON	Outstanding Notification, logically, a bitwise OR of PIR
257	SN	Suppress Notification, of non-urgent interrupts
271:258	–	Reserved (must be 0)
279:272	NV	Notification Vector, doorbell to notify about pending PIR
287:280	–	Reserved (must be 0)
319:288	NDST	Notification Destination, a physical APIC ID
511:320	–	Reserved (must be 0)

There are two additional fields in the PI descriptor that are relevant for delivering R without causing an exit. The first is the notification destination (NDST), which holds the ID of the destination LAPIC where R will fire (Core 2 in our example). The second is the notification vector (NV), which holds the value of some interrupt vector, interpreted by the destination NDST as a notification event, which triggers the direct delivery of R; the specific numerical value of NV is selected by the hypervisor (in our example, it is 242). In other words, the NV serves as a "doorbell," to let the NDST core know that some other core requested that a posted interrupt would be delivered to the guest that runs at the NDST. The hypervisor core (Core 1) sends the NV (242) to the NDST (Core 2) via an IPI (Figure 6.25b).

Observe that, in addition to the PI destination, the value of NV is also stored in the VMCS (bottom row within the gray VMCS area in Figure 6.25). As such, Core 2 understands that it has special meaning, and that it is used as a doorbell for pending posted interrupts.[8]

Importantly, if the guest is executing when the NV (242) reaches the NDST (Core 2), the handler of 242 is not invoked; without yet disturbing the guest, Core 2 needs to figure out what to do. To this end, Core 2 first appends the PIR bits into the guest's vIRR register. In so doing, it effectively turns interrupts posted by remote cores into requested interrupts in the virtual APIC of the guest. Core 2 then "evaluates pending virtual interrupts," which means that it applies the regular APIC logic with respect to the registers at the virtual APIC page. When the state of the virtual APIC permits it, the evaluation of pending virtual interrupts results in the direct delivery of an interrupt to the virtual machine.

Technically, direct virtual interrupt delivery—of an interrupt vector R—is initiated when Core 2 stores R in the "requesting virtual interrupt" (RVI) 8-bits field in the VMCS. Core 2 does that according to rules that strictly follow those of physical APICs. First, R is selected to be the maximal (highest priority) bit index set in the vIRR register. Then, R is assigned to RVI if no other interrupt vector is currently stored there, or if R's numeric value is bigger than RVI (R's priority is higher). When R resides in RVI, Core 2 evaluates whether R can actually be delivered to the guest, in which case it is moved to the VMCS "servicing virtual interrupt" (SVI) 8-bits field. Once again, the decision to deliver R is based on physical APIC rules. For example, if the guest turned off its (virtual) interrupts, or if it is currently serving a higher-priority (virtual) interrupt, R will not be delivered. When Core 2 finally assigns R to SVI and delivers R to the guest (Figure 6.25c), it likewise follows the semantics of physical interrupts, turning R off in vIRR, turning R on in vISR, and assigning the value of the highest set vIRR bit into RVI. When the guest finishes to handle R, it acknowledges the completion through its vEOI, prompting Core 2 to remove R from vISR and VSR and to initiate the evaluation procedure of pending virtual interrupts yet again.

The hypervisor at Core 2 remains uninvolved in all of this activity. It is not notified regarding any of these events unless it explicitly sets VMCS bits that declare that it would like to be notified.

IOMMU Posted Interrupts: Given that hardware support for CPU posted interrupts is available, the task of additionally supporting IOMMU posted interrupts—which directly channel interrupts from assigned devices to their guests without host involvement—is straightforward. Figure 6.26 illustrates IOMMU posted interrupts. The figure in fact encompasses and extends Figure 6.17, which depicts interrupt delivery with IOMMU interrupt remapping (IR). The main difference between the two is that, with IOMMU posted interrupt, an IR table entry may now

---

[8]At this point, readers may wonder why NV is stored twice: in the PI descriptor as well as in the VMCS. In fact, for CPU posted interrupts, the NV need not reside in the PI descriptor—the hardware is indifferent [107]. The same, by the way, applies to NDST. The presence of NV and NDST in the PI descriptor becomes crucial for IOMMU posted interrupts, as will be explained later on.

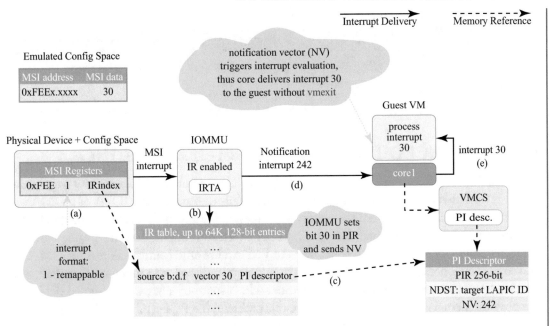

Figure 6.26: IOMMU (VT-d) posted interrupt illustration (compare with Figure 6.17).

specify the interrupt vector that the guest expects to receive when its assigned device fires an interrupt (Figure 6.17a), as well as a pointer to the PI descriptor of that guest. Upon receiving the interrupt, the IOMMU consults the associated IR table entry (Figure 6.17b). Through this entry, the IOMMU accesses the guest's PI descriptor, turns on the appropriate bit in the PIR (30 in our example), and retrieves the NV (242 in our example) and NDST (Core 1's LAPIC in our example) that are associated with the guest (Figure 6.17c). It then delivers NV to Core 1, which initiates the chain of events described above under CPU posted interrupts.

> In summary, the IOMMU and SRIOV allow for safe and scalable direct device assignment. Exitless/posted interrupts additionally enable direct interrupt delivery, making I/O virtualization performance approach bare-metal performance. The downside of direct device assignment is losing I/O interposition. The pros and cons of the alternative virtual I/O models are summarized in Table 6.4.

Table 6.4: Pros and cons of the three virtual I/O models

Virtual I/O Model	Emulation	Paravirtualization	Device Assignment
Portable, no host-specific software	✓	✗	✓
Interposition and memory optimizations	✓	✓	✗
Performance	worse	better	best

6.5   ADVANCED TOPICS AND FURTHER READING

I/O Page Faults: Recall that currently most I/O devices are unable to tolerate I/O page faults, which is the situation that arises when the IOMMU fails to translate the IOVA of a DMA operation. This inability means that the hypervisor cannot apply any of the canonical memory optimizations to VMs with assigned devices—memory overcommitment, demand-paging based on actual use, memory swapping, etc.—because the VM can use any memory address for DMAing, and so the corresponding VM image regions should physically reside in memory. Therefore, production hypervisors resort to pinning the entire address space of VMs with assigned devices to the physical memory. In an attempt to resolve this problem, the PCI-SIG recently supplemented the ATS (address translation services) PCIe standard with PRI (page request interface) [140]. ATS/PRI standardize the interaction between the I/O device, the IOMMU, and the OS, such that they are able to resolve I/O page faults. Roughly, with ATS/PRI, the IOMMU is able to tell the OS that it encountered a page fault, the OS is able to respond when the fault is resolved, and the device is able to wait until then. The standard permits I/O devices to cache translations in their own IOTLBs, and so the OS is able to tell the device to invalidate such translations when needed.

The effort to utilize ATS/PRI is spearheaded by AMD's HSA—heterogeneous system architecture [89]. HSA is aimed at unifying the address spaces of CPUs and on-die GPUs, enabling seamless page fault resolution and thereby making pinning/copying between them unnecessary [90, 119]. The HSA goals are making SOCs (which combine CPU and GPU cores) "easier to program; easier to optimize; [and provide] higher performance" [152]. GPUs process data that is *local* to the server, and they can thus be adequately served by the ATS/PRI standard—when a GPU experiences a page fault, the relevant GPU thread(s) can be easily suspended until the fault is resolved, without losing any of the data. This reasoning fails when applied to a NIC that attempts to receive incoming data, which is *external* to the server. If the corresponding DMA-write of the NIC triggers an I/O page fault, the NIC is left with no room to store the external data. More incoming data might arrive subsequently (at full line rate), and it too might fault. NICs must therefore employ some other solution to be able to tolerate I/O page faults. Such a solution is described by Lesokhin et al. [120]; it is already deployed in recent Mellanox InfiniBand NICs.

Sidecores: I/O page faults allow the hypervisor to apply the canonical memory optimizations to images of VMs that have directly assigned devices. Still, the guest I/O that flows through these assigned devices is, by definition, not interposable. As noted earlier, the superior performance associated with direct device assignment comes at a cost—the hypervisor cannot observe or manipulate the guest I/O traffic, which prevents many of the benefits of I/O interposition (§6.1). For this reason, researchers are working to develop an improved interposable virtual I/O model, whose performance is competitive to that of direct device assignment. This model makes use of hypervisor (side)cores that are dedicated to processing the virtual I/O of the VMs. Briefly, in the sidecore model, VMs do not experience virtualization exits. Instead the hypervisor sidecores poll on relevant memory locations of guest VMs, repeatedly observing the memory values and reacting accordingly. More specifically, each VM writes its I/O requests to memory shared with the host as is usual for the trap-and-emulate I/O model. Unusual, however, is that the VM does not trigger an exit. Instead, the host sidecores poll the relevant memory regions and process the request on arrival.

The benefits of the sidecores I/O model are: (1) more cycles to the VMs, whose guest OSes are asynchronous in nature and do not block-wait for I/O; (2) less cache pollution on the cores of the VMs since the hypervisor I/O-handling code no longer executes on these cores; and (3) an easily controllable, more performant scheduling of I/O threads,[9] because, in this I/O model, I/O threads run exclusively on sidecores and hence no longer interfere with the execution of VMs. The combined effect of the three benefits yields a substantial performance improvement. For example, the Elvis system—which implements paravirtual block and net devices on sidecores—is up to 3x more performant than baseline paravirtualization and is oftentimes on par with SRIOV+ELI [83]. The sidecore approach has been successfully applied to various other tasks, including IOMMU virtualization [12], storage virtualization [34], GPU processing [159], and rack-scale computing [116].

---

[9]Recall that each virtual device is implemented in a different I/O thread of the hypervisor, as depicted in Figure 6.7.

CHAPTER 7

Virtualization Support in ARM Processors

This chapter describes the current, state-of-the-art support for virtualization in modern ARM processors. Much of this work is based on the original paper on KVM/ARM, the Linux kernel virtual machine for ARM [60]. §7.1 first describes the key design principles behind ARM's virtualization technology. §7.2 describes the approach to CPU virtualization, the concept of the EL2 hypervisor mode, and how the architecture relates to the Popek and Goldberg theorem. §7.3 discusses the closely related question of MMU virtualization through nested page tables. §7.4 discusses support for interrupt virtualization. §7.5 discusses support for timer virtualization. §7.6 uses KVM/ARM as its case study on how to build a hypervisor explicitly designed to assume ARM hardware support for virtualization, and contrast it with KVM x86, as discussed in Chapter 4. §7.7 discusses some micro-architectural implications of the design, including the performance overhead of the new architectural features. §7.8 discusses the implementation complexity of KVM/ARM. §7.9 discusses improvements to the ARM virtualization support for type-2 hypervisors such as KVM/ARM. Finally, like all chapters, we close with pointers for further reading.

7.1 DESIGN PRINCIPLES OF VIRTUALIZATION SUPPORT ON ARM

ARM Architecture Virtualization Extensions [39] were introduced in 2010 as part of the ARMv7 architecture to provide architectural support for virtualization. Until then, virtualization solutions on ARM systems were based on paravirtualization and not widely used [28, 59, 91]. However, as ARM CPUs continued to increase in performance, interest in ARM virtualization also grew. ARM CPUs are pushing upward from mobile devices such as smartphones and tablets into traditional servers where virtualization is an important technology. Given ARM's widespread use in embedded systems and its business model of licensing to silicon partners who then build ARM chips, it was important for ARM to support virtualization but in a manner that enabled its partners to implement it in a cost effective manner. The central design goal of ARM Virtualization Extensions was to make it possible to build ARM hypervisors that can run unmodified guest operating systems without adding significant hardware complexity, while maintaining high levels of performance. With the success of Intel Virtualization Technology, ARM recognized the benefit of hardware virtualization support to overcome the limitation of previous ARM architectures not being classically virtualizable without the need for CPU paravirtualization or dynamic binary translation.

ARM's central design goal is to fully meet the requirements of the Popek and Goldberg theorem, with the explicit goal that virtual machines running on top of a Virtualization Extensions-based hypervisor meet the following three core attributes of equivalence, safety, and performance:

Equivalence: ARM's architects designed the Virtualization Extensions to provide architectural compatibility between the virtualized hardware and the underlying hardware, including being backward-compatible with older generations of ARM's ISA. However, ARM also introduced in ARMv7 a hardware technology for providing hardware-based security known as TrustZone. By design, TrustZone is not virtualized by the Virtualization Extensions. This view of equivalence is the same as that for Intel's VT-x from the pragmatic point of view of applications and operating systems.

Safety: ARM Virtualization Extensions also meet the safety requirement in a manner similar to Intel's VT-x. Through architectural support specifically designed for virtualization, a simpler hypervisor can provide the same characteristics with a smaller code base. This reduces the potential attack surface on the hypervisor and the risk of software vulnerabilities.

Performance: ARM benefited from the hindsight of Intel's VT-x in its design, so that delivering good performance was a goal from the beginning for first-generation hardware support for virtualization. For example, nested page tables, not part of the original x86 virtualization hardware, are standard in ARM. Furthermore, ARM went beyond Intel's VT-x in various aspects of its design to provide additional hardware mechanisms to improve virtualization performance, especially in the presence of multicore systems which have become more commonplace.

7.2   CPU VIRTUALIZATION

Like Intel's VT-x design, the architecture of the Virtualization Extensions focuses on supporting the existing ISA in the context of virtualization. ARM does not change the semantics of individual instructions of the ISA, or separately address the individual aspects of the architecture limiting virtualization, such as the non-virtualizable instructions. Instead, the Virtualization Extensions introduces a new more privileged mode of execution of the processor, referred to as **HYP** (hypervisor) mode in the ARMv7 architecture when it was first introduced, but now referred to as **EL2** since the introduction of the ARMv8 architecture.

We first provide an overview of the ARM architecture then discuss its virtualization support. Figure 7.1 shows the overall CPU modes on the ARM architecture, including **TrustZone** (Security Extensions). TrustZone splits the modes into two worlds, secure and non-secure, which are orthogonal to the CPU modes. A special mode, **monitor mode**, is provided to switch between the secure and non-secure worlds. Although ARM CPUs power up starting in the secure world, assuming TrustZone is implemented, ARM bootloaders typically transition to the non-secure world at an early stage. The secure world is used for trusted computing use cases such as digital rights management. TrustZone may appear useful for virtualization by using the secure world for hypervisor execution, but this does not work because trap-and-emulate is not supported. There is no means to trap operations executed in the non-secure world to the secure world. Non-secure software can therefore freely configure, for example, virtual memory. Any software running at the highest non-secure privilege level therefore has access to all non-secure physical memory, making it problematic to isolate multiple VMs running in the non-secure world. While it may be possible to retrofit TrustZone to support virtualization, that was not the intent of TrustZone, and it would be undesirable to exclude current uses of TrustZone in a system running hypervisors because it is being used for virtualization support.

Figure 7.1 shows a new CPU privilege level called EL2 introduced to support virtualization. ARM Virtualization Extensions are centered around this new CPU privilege level (also known as **exception level**), EL2, added to the existing user and kernel levels, EL0 and EL1, respectively. Since ARM software stacks generally run in the non-secure world, EL2 was introduced as a trap-and-emulate mechanism to support virtualization in the non-secure world. EL2 is a CPU mode that is strictly more privileged than the other CPU modes, EL0 and EL1. Note that ARMv8 only has one kernel mode, EL1, unlike previous ARM architecture versions that supported multiple kernel modes as discussed in §2.5.3.

To support VMs, software running in EL2 can configure the hardware to trap to EL2 from EL0 or EL1 on various sensitive instructions and hardware interrupts. This configurability allows for more flexible uses of the system. Consider three examples. First, a partitioning hypervisor that does not support resource overcommittment may not trap on some instructions that a normal hypervisor would trap on. Second, it may be desirable to allow a VM to use ARM debug registers directly and therefore not trap accesses to them, but if debugging on the host was being done at the same time, the VM should not be allowed to access them but should instead trap and their behavior should be emulated. Third, if one wanted to just emulate a different CPU from the one being used, it would be necessary to trap on accesses to the CPU identifier registers, but it would not be necessary to trap on other accesses required for running a full hypervisor.

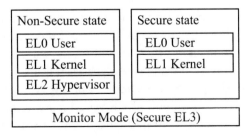

Figure 7.1: ARM processor modes.

To allow VMs to interact with an interface identical to that of the physical machine while isolating them from the rest of the system and preventing them from gaining full access to the hardware, a hypervisor enables the virtualization features in EL2 before switching to a VM. The VM will then execute normally in EL0 and EL1 until some condition is reached that requires intervention of the hypervisor. At this point, the hardware traps into EL2 giving control to the hypervisor, which can then manage the hardware and provide the required isolation across VMs. Once the condition is processed by the hypervisor, the CPU can be switched back into EL0 or EL1 and the VM can resume execution. When all virtualization features are disabled in EL2, software running in EL1 and EL0 works just like on a system without the Virtualization Extensions where software running in EL1 has full access to the hardware.

The ARM architecture allows each trap to be configured to trap directly into a VM's EL1 instead of going through EL2. For example, traps caused by system calls or page faults from EL0 can be configured to trap to a VM's EL1 directly so that they are handled by the guest OS without intervention of the hypervisor. This avoids going to EL2 on each system call or page fault, reducing virtualization overhead. Additionally, all traps into EL2 can be disabled and a single non-virtualized kernel can run in EL1 and have complete control of the system.

ARM designed the virtualization support around a separate CPU mode distinct from existing privileged modes, because they envisioned a standalone hypervisor underneath a more complex rich OS kernel [62]. They wanted to make it easier for silicon partners to implement the Virtualization Extensions, and therefore reduced the number of control registers available in EL2 compared to EL1. For example, EL2 only has one page table base register while EL1 can make use of two. Arguably, by reducing the number of control registers available in EL2, the hypervisor implementation could be made simpler by reducing the amount of state that needed to be manipulated. Similarly, they mandated certain bits to be set in the page table entries, because they did not envision a hypervisor sharing page tables with software running in user space, which is for example what the Linux kernel does with kernel mode.

Compared to Intel's VT-x, ARM Virtualization Extensions reflect a different approach with respect to CPU virtualization. ARM supports virtualization through a separate CPU mode, EL2, which is a separate and strictly more privileged CPU mode than previous user and kernel modes. In contrast, Intel has root and non-root mode, which are orthogonal to the CPU protection modes. While sensitive operations on ARM trap to EL2, sensitive operations can trap from non-root mode to root mode while staying in the same protection level on Intel. A crucial difference between the two hardware designs is that Intel's root mode supports the same full range of user and kernel mode functionality as its non-root mode, whereas ARM's EL2 is a strictly different CPU mode with its own set of features. A hypervisor using ARM's EL2 has an arguably simpler set of features to use than the more complex options available with Intel's root mode. More importantly, the simpler set of features of ARM's EL2 are most likely easier for hardware vendors to implement than the requirement of duplicating the entire architecturally visible state of the processor as done with Intel's root mode.

Both ARM and Intel trap into their respective EL2 and root modes, but Intel provides specific hardware support for a VM control block which is automatically saved and restored when switching to and from root mode using only a single instruction. This is used to automatically save and restore guest state when switching between guest and hypervisor execution. In contrast, ARM provides no such hardware support and any state that needs to be saved and restored must be done explicitly in software. The absence of this hardware support in ARM reduces the hardware complexity of implementing virtualization support, but it also provides some flexibility in what is saved and restored in switching to and from EL2. For example, trapping to ARM's EL2 is potentially faster than trapping to Intel's root mode if there is no additional state to save. On the other hand, if much of the state in the software equivalent of a VM control block needs to be saved and restored on each switch to the hypervisor, the hardware-supported VM control block feature is most likely faster. Exactly what state needs to be saved and restored when switching to and from EL2 depends on the hypervisor design and what state the hypervisor needs to use.

Compared to the properties of Intel's root mode, ARM's Virtualization Extensions provide the following properties:

- the processor is at any point in time in exactly one CPU mode, EL2, EL1, or EL0. The actual transition from one mode to another is atomic, but the process of saving and restoring state as needed when transition from one mode to another is not atomic;

- the CPU mode can be determined by executing the MRS instruction, which is available in any mode as discussed in §2.5.3. The expectation is that applications and operating systems will continue to run using the same CPU modes, whether within a VM or not. However, under this assumption, EL2 execution itself cannot be virtualized and recursive virtual machines are not supported since a nested hypervisor can determine that it is not running in EL2. This is under revision though and architecture support for nested virtualization on ARM will become available starting with ARMv8.3 [38];

- EL2 is designed to be used for virtualization, but it is simply a more privileged CPU mode and could be used for other purposes;

- each mode has its own distinct complete linear address space defined by a distinct page table. EL2 has its own **translation regime**, which defines the registers and page table formats used in a given mode. However, EL1/EL0 share a translation regime so that both of their address spaces and page tables can be accessible in either EL1 or EL0. Only entries in the TLB for the currently active address space are matched as TLB entries are each tagged with a translation regime. Since the entries are tagged, there is no need to flush the TLB in transitioning between EL2 and other CPU modes; and

- ARM has two types of interrupts—normal (IRQ) and fast (FIQ)—and two corresponding interrupt flags for the current mode of the processor. Software in EL1 can freely manipulate the interrupt flag for each type of interrupt without trapping. EL2 can configure each type of interrupt to either be delivered directly to EL1 or to trap to EL2. When the interrupt is delivered directly to EL1, the interrupt flag configured directly by EL1 controls real physical interrupts. When the interrupt traps to EL2, the interrupt flag configured directly by EL1 controls virtual interrupts.

7.2.1   VIRTUALIZATION EXTENSIONS AND THE POPEK/GOLDBERG THEOREM

Recall Popek and Goldberg's central virtualization theorem, discussed in §2.2.

> Theorem 1 [143]: For any conventional third-generation computer, a virtual machine monitor may be constructed if the set of sensitive instructions for that computer is a subset of the set of privileged instructions.

The Virtualization Extensions architecture meets the criteria of the theorem, but through a significant departure from the original model proposed to demonstrate it. ARM pragmatically took a different path by introducing a new more privileged EL2 that operates below and maintains the existing CPU modes. This ensures backward compatibility for the ISA. Terms must therefore be redefined to convince oneselves that the Virtualization Extensions follow the Popek/Goldberg criteria. The corresponding core Virtualization Extensions design principle can be informally framed as follows.

> In an architecture with an additional, separate more privileged hypervisor mode of execution, a VMM may be constructed if all sensitive instructions (according to the non-virtualizable legacy architecture) are hypervisor-mode privileged. When executing in non-hypervisor mode, all hypervisor-mode-privileged instructions are either (i) implemented by the processor, with the requirement that they operate exclusively on the non-hypervisor state of the processor, or (ii) cause a trap to the hypervisor mode.

We make two observations: (i) the theorem does not take into account whether instructions are privileged or not, but instead only takes into consideration the orthogonal question of whether and how they execute in non-hypervisor mode; and (ii) only traps are required to meet the equivalence and safety criteria. However, reducing transitions by implementing certain sensitive instructions in hardware is necessary to meet the performance criteria.

7.3   MEMORY VIRTUALIZATION

ARM and Intel are quite similar in their support for virtualizing physical memory. Both introduce an additional set of page tables to translate guest to host physical addresses, although ARM benefited from hindsight by including this feature as part of its initial virtualization support whereas Intel did not include its equivalent Extended Page Table (EPT) support, discussed in §5.1, until its second-generation virtualization hardware. Using ARM's hardware support to virtualize physical memory, when running a VM, the physical addresses managed by the VM are actually **Intermediate Physical Addresses** (IPAs), also known as guest physical addresses (gPAs), and need to be translated into host physical addresses (hPAs). Similarly to nested page tables on x86, ARM provides a second set of page tables, **Stage-2 page tables**, which translate from gPAs to hPAs corresponding to guest and host physical addresses, respectively. Stage-2 translation can be completely disabled and enabled from EL2. Stage-2 page tables use ARM's new LPAE (Large Physical Address Extension) page table format, with subtle differences from the page tables used by kernel mode.

Figure 7.2 shows the complete address translation scheme. Three levels of page tables are used for Stage-1 translation from virtual to guest physical addresses, and four levels of page tables are used for Stage-2 translation from guest to host physical addresses. Stage-2 translation can be entirely enabled or disabled using a bit in the *Hyp Configuration Register* (HCR). The base register for the Stage-2 first-level (L1) page table is specified by the *Virtualization Translation Table Base Register* (VTTBR). Both registers are only configurable from EL2.

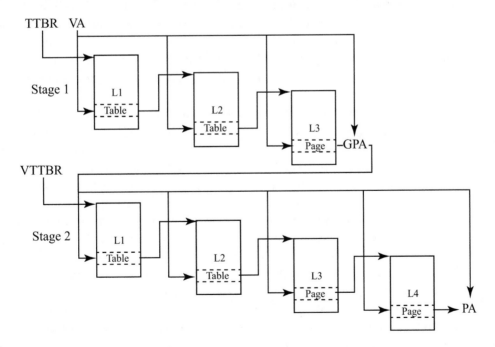

Figure 7.2: Stage-1 and Stage-2 page table walk on ARM using the LPAE memory long format descriptors. The virtual address (VA) is first translated into a guest physical address (gPA) and finally into a host physical address (hPA).

7.4   INTERRUPT VIRTUALIZATION

ARM defines the Generic Interrupt Controller (GIC) architecture [21]. The GIC routes interrupts from devices to CPUs and CPUs query the GIC to discover the source of an interrupt. The GIC is especially important in multicore configurations, because it is used to generate Inter-Processor Interrupts (IPIs) from one CPU core to another. The GIC is split in two parts, the distributor and the CPU interfaces. There is only one distributor in a system, but each CPU core has a GIC CPU interface. Both the CPU interfaces and the distributor are accessed over a Memory-Mapped interface (MMIO). The distributor is used to configure the GIC, for example, to set the CPU core affinity of an interrupt, to completely enable or disable interrupts on a system, or to send an IPI to another CPU core. The CPU interface is used to acknowledge (ACK) and to signal End-Of-Interrupt (EOI). For example, when a CPU core receives an interrupt, it will read a special register on the GIC CPU interface, which ACKs the interrupt and returns the number of the interrupt. The interrupt will not be raised to the CPU again before the CPU writes to the EOI register of the CPU interface with the value retrieved from the ACK register.

Interrupts can be configured to trap to either EL2 or EL1. Trapping all interrupts to EL1 and letting OS software running in EL1 handle them directly is efficient, but does not work in the context of VMs, because the hypervisor loses control over the hardware. Trapping all interrupts to EL2 ensures that the hypervisor retains control, but requires emulating virtual interrupts in software to signal events to VMs. This is cumbersome to manage and expensive because each step of interrupt and virtual interrupt processing, such as ACKing and EOIing, must go through the hypervisor.

The GIC includes hardware virtualization support in the form of a virtual GIC (VGIC) so that receiving virtual interrupts does not need to be emulated in software by the hypervisor. The VGIC introduces a VGIC CPU interface as well as a corresponding hypervisor control interface for each CPU. VMs are configured to see the VGIC CPU interface instead of the GIC CPU interface. Virtual interrupts are generated by writing to special registers, the *list registers*, in the VGIC hypervisor control interface, and the VGIC CPU interface raises the virtual interrupts directly to a VM's kernel mode. Because the VGIC CPU interface includes support for ACK and EOI, these operations no longer need to trap to the hypervisor to be emulated in software, reducing overhead for receiving interrupts on a CPU. For example, emulated virtual devices typically raise virtual interrupts through a software API to the hypervisor, which can leverage the VGIC by writing the virtual interrupt number for the emulated device into the list registers. This causes the VGIC to interrupt the VM directly to kernel mode and lets the guest OS ACK and EOI the virtual interrupt without trapping to the hypervisor. Note that the distributor must still be emulated in software and all accesses to the distributor by a VM must still trap to the hypervisor. For example, when a virtual CPU sends a virtual IPI to another virtual CPU, this will cause a trap to the hypervisor, which emulates the distributor access in software and programs the list registers on the receiving CPU's GIC hypervisor control interface.

ARM's support for virtual interrupts had no x86 counterpart until the more recent introduction of Intel's virtual APIC support [104], as discussed in Chapter 6. Similar to ARM's support for virtual interrupts, Intel's APIC virtualization support also allows VMs to EOI interrupts without trapping to the hypervisor. Furthermore, the Intel virtual APIC support allows VMs to access a number of APIC registers without trapping to the hypervisor by providing a backing page in memory for the VM's virtualized APIC state. While Intel has recently introduced posted interrupts, x86 support for direct interrupt delivery, ARM has introduced a similar mechanism in version 4.0 of the GIC architecture (GICv4) [22]. However, GICv4 is not yet available in current ARM hardware.

7.5   TIMER VIRTUALIZATION

ARM defines the Generic Timer Architecture which includes support for timer virtualization. Generic timers provide a *counter* that measures passing of time in real-time, and a *timer* for each CPU, which is programmed to raise an interrupt to the CPU after a certain amount of time has passed. Timers are likely to be used by both hypervisors and guest OSes, but to provide isolation and retain control, the timers used by the hypervisor cannot be directly configured and manipulated by guest OSes. Such timer accesses from a guest OS would need to trap to EL2, incurring additional overhead for a relatively frequent operation for some workloads. Hypervisors may also wish to virtualize VM time, which is problematic if VMs have direct access to counter hardware.

ARM provides virtualization support for the timers by introducing a new counter, the *virtual counter* and a new timer, the *virtual timer*. A hypervisor can be configured to use physical timers while VMs are configured to use virtual timers. VMs can then access, program, and cancel virtual timers without causing traps to EL2. Access to the physical timer and counter from kernel mode is controlled from EL2, but software running in kernel mode always has access to the virtual timers and counters. Additionally, EL2 configures an offset register, which is subtracted from the physical counter and returned as the value when reading the virtual counter. Note that prior to the use of generic timers, the frequent operation of simply reading a counter is a memory mapped operation, which would typically trap to EL2 and generate additional overhead.

ARM's support for virtual timers has no real x86 counterpart. The x86 world of timekeeping consists of a myriad of timing devices available partially due to the history and legacy of the PC platform. Modern x86 platforms typically support an 8250-series Programmable Interval Timer (PIT) for legacy support and a Local APIC (LAPIC) timer. The Intel hardware support for virtualization adds the VMX-Preemption timer which allows hypervisors to program an exit from a VM independently from how other timers are programmed. The VMX-Preemption timer was added to reduce the latency between a timer firing and the hypervisor injecting a virtual timer interrupt. This is achieved because a hypervisor doesn't have to handle an interrupt from the LAPIC timer, but can directly tell from a VM exit that the preemption timer has expired. Contrary to ARM, x86 does not support giving full control of individual timer hardware to VMs.

x86 does allow VMs to directly read the Time Stamp Counter (TSC), wheres ARM allows access to the virtual counter. The x86 TSC is typically higher resolution than the ARM counter, because the TSC is driven by the processor's clock where the ARM counter is driven by a dedicated clock signal.

7.6 KVM/ARM—A VMM BASED ON ARM VIRTUALIZATION EXTENSIONS

So far, we have described the hardware extensions introduced by ARM's Virtualization Extensions, and discussed a few architectural considerations. We now use KVM [113], the Linux-based Kernel Virtual Machine, as a case study of how to adapt its design for ARM from its original design for VT-x to create KVM/ARM, the Linux ARM hypervisor [60].

Instead of reinventing and reimplementing complex core functionality in the hypervisor, and potentially introducing tricky and fatal bugs along the way, KVM/ARM builds on KVM and leverages existing infrastructure in the Linux kernel. While a standalone bare metal hypervisor design approach has the potential for better performance and a smaller Trusted Computing Base (TCB), this approach is less practical on ARM. ARM hardware is in many ways much more diverse than x86. Hardware components are often tightly integrated in ARM devices in non-standard ways by different device manufacturers. ARM hardware lacks features for hardware discovery such as a standard BIOS or a PCI bus, and there is no established mechanism for installing low-level software on a wide variety of ARM platforms. Linux, however, is supported across almost all ARM platforms and by integrating KVM/ARM with Linux, KVM/ARM is automatically available on any device running a recent version of the Linux kernel. This is in contrast to bare metal approaches such as Xen [188], which must actively support every platform on which they wish to install the Xen hypervisor. For example, for every new SoC that Xen needs to support, the developers must implement a new serial device driver in the core Xen hypervisor.

While KVM/ARM benefits from its integration with Linux in terms of portability and hardware support, a key problem that needed to be addressed is that the ARM hardware virtualization extensions were designed to support a standalone hypervisor design where the hypervisor is completely separate from any standard kernel functionality. In the following, we describe how KVM/ARM's design makes it possible to benefit from integration with an existing kernel and at the same time take advantage of the hardware virtualization features.

7.6.1 SPLIT-MODE VIRTUALIZATION

Simply running a hypervisor entirely in ARM's EL2 is attractive since it is the most privileged level. However, since KVM/ARM leverages existing kernel infrastructure such as the scheduler, running KVM/ARM in EL2 implies running the Linux kernel in EL2. This is problematic for at least two reasons. First, low-level architecture dependent code in Linux is written to work in kernel mode, and would not run unmodified in EL2, because EL2 is a completely different CPU mode from normal kernel mode. The significant changes required to run the kernel in EL2 would be very unlikely to be accepted by the Linux kernel community. More importantly, to preserve compatibility with hardware without EL2 and to run Linux as a guest OS, low-level code would have to be written to work in both modes, potentially resulting in slow and convoluted code paths. As a simple example, a page fault handler needs to obtain the virtual address causing the page fault. In EL2, this address is stored in a different register than in kernel mode.

Second, running the entire kernel in EL2 would adversely affect native performance. For example, EL2 has its own separate address space. Whereas kernel mode uses two page table base registers to provide the familiar 3 GB/1 GB split between user address space and kernel address space, EL2 uses a single page table register and therefore cannot have direct access to the user space portion of the address space. Frequently used functions to access user memory would require the kernel to explicitly map user space data into kernel address space and subsequently perform necessary teardown and TLB maintenance operations, resulting in poor native performance on ARM.

These problems with running a Linux hypervisor using ARM EL2 do not occur for x86 hardware virtualization. x86 root mode is orthogonal to its CPU privilege modes. The entire Linux kernel can run in root mode as a hypervisor because the same set of CPU modes available in non-root mode are available in root mode. Nevertheless, given the widespread use of ARM and the advantages of Linux on ARM, finding an efficient virtualization solution for ARM that can leverage Linux and take advantage of the hardware virtualization support is of crucial importance.

KVM/ARM introduces **split-mode virtualization**, a new approach to hypervisor design that splits the core hypervisor so that it runs across different privileged CPU modes to take advantage of the specific benefits and functionality offered by each CPU mode. KVM/ARM uses split-mode virtualization to leverage the ARM hardware virtualization support enabled by EL2, while at the same time leveraging existing Linux kernel services running in kernel mode. Split-mode virtualization allows KVM/ARM to be integrated with the Linux kernel without intrusive modifications to the existing code base.

This is done by splitting the hypervisor into two components, the **lowvisor** and the **highvisor**, as shown in Figure 7.3. The lowvisor is designed to take advantage of the hardware virtualization support available in EL2 to provide three key functions. First, the lowvisor sets up the correct execution context by appropriate configuration of the hardware, and enforces protection and isolation between different execution contexts. The lowvisor directly interacts with hardware protection features and is therefore highly critical and the code base is kept to an absolute minimum. Second, the lowvisor switches from a VM execution context to the host execution context and vice-versa. The host execution context is used to run the hypervisor and the host Linux kernel. We refer to an execution context as a world, and switching from one world to another as a *world switch*, because the entire state of the system is changed. Since the lowvisor is the only component that runs in EL2, only it can be responsible for the hardware reconfiguration necessary to perform a world switch. Third, the lowvisor provides a virtualization trap handler, which handles

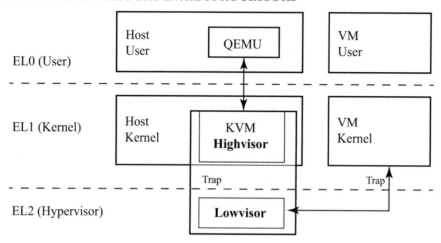

Figure 7.3: KVM/ARM system architecture.

interrupts and exceptions that must trap to the hypervisor. All traps to the hypervisor must first go to the lowvisor. The lowvisor performs only the minimal amount of processing required and defers the bulk of the work to be done to the highvisor after a world switch to the highvisor is complete.

The highvisor runs in kernel mode as part of the host Linux kernel. It can therefore directly leverage existing Linux functionality such as the scheduler, and can make use of standard kernel software data structures and mechanisms to implement its functionality, such as locking mechanisms and memory allocation functions. This makes higher-level functionality easier to implement in the highvisor. For example, while the lowvisor provides a low-level trap-handler and the low-level mechanism to switch from one world to another, the highvisor handles Stage-2 page faults from the VM and performs instruction emulation. Note that parts of the VM run in kernel mode, just like the highvisor, but with Stage-2 translation and trapping to EL2 enabled.

Because the hypervisor is split across kernel mode and EL2, switching between a VM and the highvisor involves multiple mode transitions. A trap to the highvisor while running the VM will first trap to the lowvisor running in EL2. The lowvisor will then cause another trap to run the highvisor. Similarly, going from the highvisor to a VM requires trapping from kernel mode to EL2, and then switching to the VM. As a result, split-mode virtualization incurs a double trap cost in switching to and from the highvisor. On ARM, the only way to perform these mode transitions to and from EL2 is by trapping. However, it turns out that this extra trap is not a significant performance cost on ARM, as discussed in §7.7.

KVM/ARM uses a memory mapped interface to share data between the highvisor and lowvisor as necessary. Because memory management can be complex, it leverages the highvisor's ability to use the existing memory management subsystem in Linux to manage memory for both the highvisor and lowvisor. Managing the lowvisor's memory involves additional challenges though, because it requires managing EL2's separate address space. One simplistic approach would be to reuse the host kernel's page tables and also use them in EL2 to make the address spaces identical. This unfortunately does not work, because EL2 uses a different page table format from kernel mode. Therefore, the highvisor explicitly manages the EL2 page tables to map any code executed in EL2 and any data structures shared between the highvisor and the lowvisor to the same virtual addresses in EL2 and in kernel mode.

7.6.2 CPU VIRTUALIZATION

To virtualize the CPU, KVM/ARM must present an interface to the VM which is essentially identical to the underlying real hardware CPU, while ensuring that the hypervisor remains in control of the hardware. This involves ensuring that software running in the VM must have persistent access to the same register state as software running on the physical CPU, as well as ensuring that physical hardware state associated with the hypervisor and its host kernel is persistent across running VMs. Register state not affecting VM isolation can simply be context switched by saving the VM state and restoring the host state from memory when switching from a VM to the host and vice versa. KVM/ARM configures access to all other sensitive state to trap to EL2, so it can be emulated by the hypervisor.

Table 7.1 shows the CPU register state visible to software running in kernel and user mode, and KVM/ARM's virtualization method for each register group. The lowvisor has its own dedicated configuration registers only for use in EL2, and is not shown in Table 7.1. KVM/ARM context switches registers during world-switches whenever the hardware supports it, because it allows the VM direct access to the hardware. For example, the VM can directly program the Stage-1 page table base register without trapping to the hypervisor, a fairly common operation in most guest OSes. KVM/ARM performs trap and emulate on sensitive instructions and when accessing hardware state that could affect the hypervisor or would leak information about the hardware to the VM that violates its virtualized abstraction. For example, KVM/ARM traps if a VM executes the WFI instruction, which causes the CPU to power down, because such an operation should only be performed by the hypervisor to maintain control of the hardware. KVM/ARM defers switching certain register state until absolutely necessary, which slightly improves performance under certain workloads.

The difference between running inside a VM in kernel or user mode and running the hypervisor in kernel or user mode is determined by how the virtualization extensions have been configured by EL2 during the world switch. A world switch from the host to a VM performs the following actions:

1. store all host GP registers on the EL2 stack;

2. configure the VGIC for the VM;

3. configure the timers for the VM;

Table 7.1: VM and Host State on a Cortex-A15

Action	Nr.	State
Context Switch	38	General Purpose (GP) Registers
	26	Control Registers
	16	VGIC Control Registers
	4	VGIC List Registers
	2	Arch. Timer Control Registers
	32	64-bit VFP Registers
	4	32-bit VFP Control Registers
Trap-and-Emulate	-	CP14 Trace Registers
	-	WFI Instructions
	-	SMC Instructions
	-	ACTLR Access
	-	Cache Ops. by Set/Way
	-	L2CTLR / L2ECTLR Registers

4. save all host-specific configuration registers onto the EL2 stack;

5. load the VM's configuration registers onto the hardware, which can be done without affecting current execution, because EL2 uses its own configuration registers, separate from the host state;

6. configure EL2 to trap floating-point operations for lazy context switching of floating-point (VFP) registers, trap interrupts, trap CPU halt instructions (WFI/WFE), trap SMC instructions, trap specific configuration register accesses, and trap debug register accesses;

7. write VM-specific IDs into shadow ID registers, defined by the ARM Virtualization Extensions and accessed by the VM in lieu of the hardware values in the ID registers;

8. set the Stage-2 page table base register (VTTBR) and enable Stage-2 address translation;

9. restore all guest GP registers; and

10. trap into either user or kernel mode.

The CPU will stay in the VM world until an event occurs, which triggers a trap into EL2. Such an event can be caused by any of the traps mentioned above, a Stage-2 page fault, or a hardware interrupt. Since the event requires services from the highvisor, either to emulate the expected hardware behavior for the VM or to service a device interrupt, KVM/ARM must perform another world switch back into the highvisor and its host. This entails trapping first to the lowvisor before going to the highvisor. The world switch back to the host from a VM performs the following actions:

1. store all VM GP registers;

2. disable Stage-2 translation;

3. configure EL2 to not trap any register access or instructions;

4. save all VM-specific configuration registers;

5. load the host's configuration registers onto the hardware;

6. configure the timers for the host;

7. save VM-specific VGIC state;

8. restore all host GP registers; and

9. trap into kernel mode.

7.6.3   MEMORY VIRTUALIZATION

As discussed in §7.3, ARM provides Stage-2 page tables to translate guest to host physical addresses. KVM/ARM provides memory virtualization by enabling Stage-2 translation for all memory accesses when running in a VM. Stage-2 translation can only be configured in EL2, and its use is completely transparent to the VM. The highvisor manages the Stage-2 translation page tables to only allow access to memory specifically allocated for a VM; other accesses will cause Stage-2 page faults which trap to the hypervisor. This mechanism ensures that a VM cannot access memory belonging to the hypervisor or other VMs, including any sensitive data. Stage-2 translation is disabled when running in the highvisor and lowvisor because the highvisor has full control of the complete system and directly manages the host physical addresses. When the hypervisor performs a world switch to a VM, it enables Stage-2 translation and configures the Stage-2 page table base register accordingly. Although both the highvisor and VMs share the same CPU modes, Stage-2 translations ensure that the highvisor is protected from any access by the VMs.

KVM/ARM uses split-mode virtualization to leverage existing kernel memory allocation, page reference counting, and page table manipulation code. KVM/ARM handles Stage-2 page faults by considering the gPA of the fault, and if that address belongs to normal memory in the VM memory map, KVM/ARM allocates a page for the VM by simply calling an existing kernel function, such as get_user_pages, and maps the allocated page to the VM in the Stage-2 page tables. In comparison, a bare metal hypervisor would be forced to either statically allocate memory to VMs or write an entire new memory allocation subsystem.

7.6.4   I/O VIRTUALIZATION

KVM/ARM leverages existing QEMU and Virtio [154] user space device emulation to provide I/O virtualization. At a hardware level, all I/O mechanisms on the ARM architecture are based on load/store operations to MMIO device regions. With the exception of devices directly assigned to VMs, all hardware MMIO regions are inaccessible from VMs. KVM/ARM uses Stage-2 translations to ensure that physical devices cannot be accessed directly from VMs. Any access outside of RAM regions allocated for the VM will trap to the hypervisor, which can route the access to a specific emulated device in QEMU based on the fault address. This is somewhat different from x86, which uses x86-specific hardware instructions such as `inl` and `outl` for port I/O operations in addition to MMIO. Chapter 6 provides a more in-depth discussion regarding I/O virtualization in general.

7.6.5   INTERRUPT VIRTUALIZATION

KVM/ARM leverages its tight integration with Linux to reuse existing device drivers and related functionality, including handling interrupts. When running in a VM, KVM/ARM configures the CPU to trap all hardware interrupts to EL2. On each interrupt, it performs a world switch to the highvisor and the host handles the interrupt, so that the hypervisor remains in complete control of hardware resources. When running in the host and the highvisor, interrupts trap directly to kernel mode, avoiding the overhead of going through EL2. In both cases, all hardware interrupt processing is done in the host by reusing Linux's existing interrupt handling functionality.

However, VMs must receive notifications in the form of virtual interrupts from emulated devices and multicore guest OSes must be able to send virtual IPIs from one virtual core to another. KVM/ARM uses the VGIC to inject virtual interrupts into VMs to reduce the number of traps to EL2. As described in §7.4, virtual interrupts are raised to virtual CPUs by programming the list registers in the VGIC hypervisor CPU control interface. KVM/ARM configures the Stage-2 page tables to prevent VMs from accessing the control interface and to allow access only to the VGIC virtual CPU interface, ensuring that only the hypervisor can program the control interface and that the VM can access the VGIC virtual CPU interface directly. However, guest OSes will still attempt to access a GIC distributor to configure the GIC and to send IPIs from one virtual core to another. Such accesses will trap to the hypervisor and the hypervisor must emulate the distributor.

KVM/ARM introduces the virtual distributor, a software model of the GIC distributor as part of the highvisor. The virtual distributor exposes an interface to user space, so emulated devices in user space can raise virtual interrupts to the virtual distributor, and exposes an MMIO interface to the VM identical to that of the physical GIC distributor. The virtual distributor keeps internal software state about the state of each interrupt and uses this state whenever a VM is scheduled, to program the list registers to inject virtual interrupts. For example, if virtual CPU0 sends an IPI to virtual CPU1, the distributor will program the list registers for virtual CPU1 to raise a virtual IPI interrupt the next time virtual CPU1 runs.

Ideally, the virtual distributor only accesses the hardware list registers when necessary, since device MMIO operations are typically significantly slower than cached memory accesses. A complete context switch of the list registers is required when scheduling a different VM to run on a physical core, but not necessarily required when simply switching between a VM and the hypervisor. For example, if there are no pending virtual interrupts, it is not necessary to access any of the list registers. Note that once the hypervisor writes a virtual interrupt to a list register when switching to a VM, it must also read the list register back when switching back to the hypervisor, because the list register describes the state of the virtual interrupt and indicates, for example, if the VM has ACKed the virtual interrupt. The initial unoptimized version of KVM/ARM uses a simplified approach which completely context switches all VGIC state including the list registers on each world switch.

7.6.6   TIMER VIRTUALIZATION

Reading counters and programming timers are frequent operations in many OSes for process scheduling and to regularly poll device state. For example, Linux reads a counter to determine if a process has expired its time slice, and programs timers to ensure that processes don't exceed their allowed time slices. Application workloads also often leverage timers for various reasons. Trapping to the hypervisor for each such operation is likely to incur noticeable performance overheads, and allowing a VM direct access to the time-keeping hardware typically implies giving up timing control of the hardware resources as VMs can disable timers and control the CPU for extended periods of time.

KVM/ARM leverages ARM's hardware virtualization features of the generic timers to allow VMs direct access to reading counters and programming timers without trapping to EL2 while at the same time ensuring the hypervisor remains in control of the hardware. Since access to the physical timers is controlled using EL2, any software controlling EL2 mode has access to the physical timers. KVM/ARM maintains hardware control by using the physical timers in the hypervisor and disallowing access to physical timers from the VM. The Linux kernel running as a guest OS only accesses the virtual timer and can therefore directly access timer hardware without trapping to the hypervisor.

Unfortunately, due to architectural limitations, the virtual timers cannot directly raise virtual interrupts, but always raise hardware interrupts, which trap to the hypervisor. KVM/ARM detects when a VM virtual timer expires, and injects a corresponding virtual interrupt to the VM, performing all hardware ACK and EOI operations in the highvisor. The hardware only provides a single virtual timer per physical CPU, and multiple virtual CPUs may be multiplexed across this single hardware instance. To support virtual timers in this scenario, KVM/ARM detects unexpired timers when a VM traps to the hypervisor and leverages existing OS functionality to program a software timer at the time when the virtual timer would have otherwise fired, had the VM been left running. When such a software timer fires, a callback function is executed, which raises a virtual timer interrupt to the VM using the virtual distributor described above.

7.7   PERFORMANCE MEASUREMENTS

We present some experimental results that quantify the performance of ARM virtualization by comparing the performance of both x86 and ARM implementations of KVM on multicore hardware. ARM measurements were obtained using an Insignal Arndale board [97] with a dual core 1.7 GHz Cortex A-15 CPU on a Samsung Exynos 5250 SoC. This is the first and most widely used commercially available development board based on the Cortex A-15, the first ARM CPU with hardware virtualization support. Onboard 100 Mb Ethernet is provided via the USB bus and an external 120 GB Samsung 840 series SSD drive was connected to the Arndale board via eSATA. x86 measurements were obtained using both a low-power mobile laptop platform and an industry standard server platform. The laptop platform was a 2011 MacBook Air with a dual core 1.8 GHz Core i7-2677M CPU, an internal Samsung SM256C 256 GB SSD drive, and an Apple 100 Mb USB Ethernet adapter. The server platform was a dedicated OVH SP 3 server with a dual core 3.4 GHz Intel Xeon E3 1245v2 CPU, two physical SSD drives of which only one was used, and 1 Gb Ethernet dropped down to connect to a 100 Mb network infrastructure.

To provide comparable measurements, the software environments across all hardware platforms were kept the same as much as possible. Both the host and guest VMs on all platforms were Ubuntu version 12.10. The mainline Linux 3.10 kernel was used for the experiments, with patches for huge page support applied on top of the source tree. Since the experiments were performed on a number of different platforms, the kernel configurations had to be slightly different, but all common features were configured similarly across all platforms. In particular, Virtio drivers were used in the guest VMs on both ARM and x86. QEMU version v1.5.0 was used for the measurements. All systems were configured with a maximum of 1.5 GB of RAM available to the respective guest VM or host being tested. Furthermore, all multicore measurements were done using two physical cores and guest VMs with two virtual CPUs, and single-core measurements were configured with SMP disabled in the kernel configuration of both the guest and host system; hyperthreading was disabled on the x86 platforms. CPU frequency scaling was disabled to ensure that native and virtualized performance was measured at the same clock rate on each platform.

Table 7.2 presents various micro-architectural costs of virtualization using KVM/ARM on ARM and KVM x86 on x86. The measurements were obtained using custom small guest OSes [57, 112] with some bugfix patches applied. Code for both KVM/ARM and KVM x86 was instrumented to read the cycle counter at specific points along critical paths to more accurately determine where overhead time was spent. Measurements are shown in cycles instead of time to provide a useful comparison across platforms with different CPU frequencies. We show two numbers for the ARM platform where possible, with and without VGIC and virtual timers support.

Hypercall is the cost of two world switches, going from the VM to the host and immediately back again without doing any work in the host. KVM/ARM takes three to four times as many cycles for this operation vs. KVM x86 due to two main factors. First, saving and restoring VGIC state to use virtual interrupts is quite expensive on ARM. The ARM without VGIC/vtimers

Table 7.2: Micro-architectural cycle counts

Micro Test	ARM	ARM no VGIC/vtimers	x86 Laptop	x86 Server
Hypercall	5.326	2,270	1,336	1,638
Trap	27	27	632	821
IPI	14,366	32,951	17,138	21,177
EOI+ACK	427	13,726	2,043	2,305

measurement does not include the cost of saving and restoring VGIC state, showing that this accounts for over half of the cost of a world switch on ARM. Second, x86 provides hardware support to save and restore state on the world switch, which is much faster. ARM requires software to explicitly save and restore state, which provides greater flexibility, but higher costs.

Trap is the cost of switching the hardware mode from the VM into the respective CPU mode for running the hypervisor, EL2 on ARM and root mode on x86. ARM is much faster than x86 because it only needs to manipulate two registers to perform this trap, whereas the cost of a trap on x86 is roughly the same as the cost of a world switch because the same amount of state is saved by the hardware in both cases. The trap cost on ARM is a very small part of the world switch costs, indicating that the double trap incurred by split-mode virtualization on ARM does not add much overhead.

That is not to say that the cost of split-mode virtualization is necessarily small, as the double trap is a small part of the overall hypercall cost. There are three other more substantial costs involved. First, because the host OS and the VM both run in EL1 and ARM hardware does not provide any features to distinguish between the host OS running in EL1 and the VM running in EL1, software running in EL2 must context switch all the EL1 system register state between the VM guest OS and the type-2 hypervisor host OS, incurring added cost of saving and restoring EL1 register state. Second, because the host OS runs in EL1 and needs full access to the hardware, the hypervisor must disable traps to EL2 and Stage-2 translation from EL2 while switching from the VM to the hypervisor, and enable them when switching back to the VM again. Third, because the type-2 hypervisor runs in EL1 but needs to access VM control register state such as the VGIC state, which can only be accessed from EL2, there is additional overhead to read and write the VM control register state in EL2. The type-2 hypervisor can either jump back and forth between EL1 and EL2 to access the control register state when needed, or it can copy the full register state to memory while it is still in EL2, return to the host OS in EL1 and read and write the memory copy of the VM control state, and then finally copy the state from memory back to the EL2 control registers when the hypervisor is running in EL2 again. Both methods incur much overhead, but jumping back and forward between EL1 and EL2 makes the software implementation complicated and difficult to maintain. Therefore, the KVM/ARM implementation currently takes the second approach of reading and writing all VM control registers in EL2 during each transition between the VM and the hypervisor.

IPI is the cost of issuing an IPI to another virtual CPU core when both virtual cores are running on separate physical cores and both are actively running inside the VM. IPI measures time starting from sending an IPI until the other virtual core responds and completes the IPI. It involves multiple world switches and sending and receiving a hardware IPI. Despite its higher world switch cost, ARM is faster than x86 because the underlying hardware IPI on x86 is expensive, x86 APIC MMIO operations require KVM x86 to perform instruction decoding not needed on ARM, and completing an interrupt on x86 is more expensive. ARM without VGIC/vtimers is significantly slower than with VGIC/vtimers even though it has lower world switch costs because sending, EOIing and ACKing interrupts trap to the hypervisor and are handled by QEMU in user space.

EOI+ACK is the cost of completing a virtual interrupt on both platforms. It includes both interrupt acknowledgment and completion on ARM, but only completion on the x86 platform. ARM requires an additional operation, the acknowledgment, to the interrupt controller to determine the source of the interrupt. x86 does not have the same requirement because the source is directly indicated by the interrupt descriptor table entry at the time when the interrupt is raised. However, the operation is roughly 5 times faster on ARM than x86 because there is no need to trap to the hypervisor on ARM because of VGIC support for both operations. On x86, the EOI operation must be emulated and therefore causes a trap to the hypervisor. This operation is required for every virtual interrupt including both virtual IPIs and interrupts from virtual devices.

7.8   IMPLEMENTATION COMPLEXITY

We compare the code complexity of KVM/ARM to its KVM x86 counterpart in Linux 3.10 using cloc [61] to count lines of code (LOC). KVM/ARM is 5,812 LOC, counting just the architecture-specific code added to Linux to implement it, of which the lowvisor is a mere 718 LOC. As a conservative comparison, KVM x86 is 25,367 LOC, excluding guest performance monitoring support, not yet supported by KVM/ARM, and 3,311 LOC required for AMD support. These numbers do not include KVM's architecture-generic code, 7,071 LOC, which is shared by all systems. Table 7.3 shows a breakdown of the total hypervisor architecture-specific code into its major components.

By inspecting the code we notice the striking additional complexity in the x86 implementation is mainly due to the five following reasons.

1. Since EPT was not supported in earlier hardware versions, KVM x86 must support shadow page tables.

2. The hardware virtualization support has evolved over time, requiring software to conditionally check for support for a large number of features such as EPT.

Table 7.3: Code Complexity in Lines of Code (LOC)

Component	KVM ARM	KVM x86 (Intel)
Core CPU	2,493	16,177
Page Fault Handling	738	3,410
Interrupts	1,057	1,978
Timers	180	573
Other	1,344	1,288
Architecture-specific	5,812	25,367

3. A number of operations require software decoding of instructions on the x86 platform. KVM/ARM's out-of-tree MMIO instruction decode implementation was much simpler, only 462 LOC.

4. The various paging modes on x86 requires more software logic to handle page faults.

5. x86 requires more software logic to support interrupts and timers than ARM, which provides VGIC/vtimers hardware support that reduces software complexity.

KVM/ARM's LOC is less than partially complete bare-metal microvisors written for EL2 [173], with the lowvisor LOC almost an order of magnitude smaller. Unlike standalone hypervisors, KVM/ARM's code complexity is so small because lots of functionality simply does not have to be implemented as it is already provided by Linux. Table 7.3 does not include other non-hypervisor architecture-specific Linux code, such as basic bootstrapping, which is significantly more code. Porting a standalone hypervisor such as Xen from x86 to ARM is much more complicated because all of the ARM code for basic system functionality needs to be written from scratch. In contrast, since Linux is dominant on ARM, KVM/ARM just leverages existing Linux ARM support to run on every platform supported by Linux.

7.9   ARCHITECTURE IMPROVEMENTS

Building on the experiences of the developers of KVM/ARM, a set of improvements have been made to the ARM architecture to avoid the need for split-mode virtualization for type-2 hypervisors such as KVM/ARM. These improvements are the **Virtualization Host Extensions** (VHE), which are now part of a new revision of the ARM 64-bit architecture, ARMv8.1 [40]. VHE allows running an OS designed to run in EL1 to run in EL2 without substantial modification to the OS source code. We show how this allows KVM/ARM and its Linux host kernel to run entirely in EL2 without substantial modifications to Linux.

VHE is provided through the addition of a new control bit, the E2H bit, which is set at system boot when installing a type-2 hypervisor that uses VHE. If the bit is not set, ARMv8.1

behaves the same as ARMv8 in terms of hardware virtualization support, preserving backward compatibility with existing hypervisors. When the bit is set, VHE enables three main features.

First, VHE expands EL2, adding additional physical register state to the CPU, such that any register and functionality available in EL1 is also available in EL2. For example, EL1 has two registers, TTBR0_EL1 and TTBR1_EL1, the first used to lookup the page tables for virtual addresses (VAs) in the lower VA range, and the second in the upper VA range. This provides a convenient and efficient method for splitting the VA space between userspace and the kernel. However, without VHE, EL2 only has one page table base register, TTBR0_EL2, making it problematic to support the split VA space of EL1 when running in EL2. With VHE, EL2 gets a second page table base register, TTBR1_EL2, making it possible to support split VA space in EL2 in the same way as provided in EL1. This enables a type-2 hypervisor integrated with a host OS to support a split VA space in EL2, which is necessary to run the host OS in EL2 so it can manage the VA space between userspace and the kernel.

Second, VHE provides a mechanism to access the extra EL2 register state transparently. Simply providing extra EL2 registers is not sufficient to run unmodified OSes in EL2, because existing OSes are written to access EL1 registers. For example, Linux is written to use TTBR1_EL1, which does not affect the translation system running in EL2. Providing the additional register TTBR1_EL2 would still require modifying Linux to use the TTBR1_EL2 instead of the TTBR1_EL1 when running in EL2 vs. EL1, respectively. To avoid forcing OS vendors to add this extra level of complexity to the software, VHE allows unmodified software to execute in EL2 and transparently access EL2 registers using the EL1 register access function instruction encodings. For example, current OS software reads the TTBR1_EL1 register with the instruction `mrs x1, ttbr1_el1`. With VHE, the software still executes the same instruction, but the hardware actually accesses the TTBR1_EL2 register. As long as the E2H bit is set, accesses to EL1 registers performed in EL2, actually access EL2 registers, thereby transparently rewriting register accesses to EL2, as described above. A new set of special instructions are added to access the EL1 registers in EL2, which the hypervisor can use to switch between VMs, which will run in EL1. For example, if the hypervisor wishes to access the guest's TTBR1_EL1, it will use the instruction `mrs x1, ttb1_el21`.

Third, VHE expands the memory translation capabilities of EL2. In ARMv8, EL2 and EL1 use different page table formats so that software written to run in EL1 must be modified to run in EL2. In ARMv8.1, the EL2 page table format is now compatible with the EL1 format when the E2H bit is set. As a result, an OS that was previously run in EL1 can now run in EL2 without being modified because it can use the same EL1 page table format.

Figure 7.4 shows how type-1 and type-2 hypervisors map to the architecture with VHE. Type-1 hypervisors such as Xen, which can be designed explicitly to run in EL2 already without any additional support, do not set the E2H bit introduced with VHE, and EL2 behaves exactly as in ARMv8. Type-2 hypervisors such as KVM/ARM set the E2H bit when the system boots, and the host OS kernel runs exclusively in EL2, and never in EL1. The type-2 hypervisor kernel

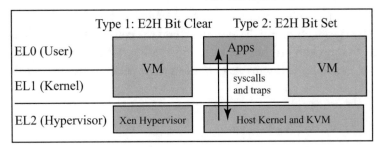

Figure 7.4: Virtualization Host Extensions (VHE).

can run unmodified in EL2, because VHE provides an equivalent EL2 register for every EL1 register and transparently rewrites EL1 register accesses from EL2 to EL2 register accesses, and because the page table formats between EL1 and EL2 are now compatible. Transitions from host userspace to host kernel happen directly from EL0 to EL2, for example to handle a system call, as indicated by the arrows in Figure 7.4. Transitions from the VM to the hypervisor now happen without having to context switch EL1 state, because EL1 is not used by the hypervisor.

ARMv8.1 differs from the x86 approach in two key ways. First, ARMv8.1 introduces more additional hardware state so that a VM running in EL1 does not need to save a substantial amount of state before switching to running the hypervisor in EL2 because the EL2 state is separate and backed by additional hardware registers. This minimizes the cost of VM to hypervisor transitions because trapping from EL1 to EL2 does not require saving and restoring state beyond general purpose registers to and from memory. In contrast, recall that the x86 approach adds CPU virtualization support by adding root and non-root mode as orthogonal concepts from the CPU privilege modes, but does not introduce additional hardware register state like ARM. As a result, switching between root and non-root modes requires transferring state between hardware registers and memory. The cost of this is ameliorated by implementing the state transfer in hardware, but while this avoids the need for additional instruction fetch and decode, accessing memory is still expected to be more expensive than having extra hardware register state. Second, ARMv8.1 preserves the RISC-style approach of allowing software more fine-grained control over which state needs to be switched for which purposes instead of fixing this in hardware, potentially making it possible to build hypervisors with lower overhead, compared to x86.

A type-2 hypervisor originally designed for ARMv8 must be modified to benefit from VHE. A patch set has been developed to add VHE support to KVM/ARM. This involves rewriting parts of the code to allow run-time adaptations of the hypervisor, such that the same kernel binary can run on both legacy ARMv8 hardware and benefit from VHE-enabled ARMv8.1 hardware. It is unfortunate that the ARM hardware support for type-2 hypervisors was not included in the original version of the virtualization extensions, because the KVM/ARM implementation now has to support both legacy ARMv8 virtualization support using split-mode virtualization and VHE in ARMv8.1 To make matters worse, this has to be a runtime decision, because the same kernel binary must be able to run across both ARMv8 and ARMv8.1 server hardware, both of which are in active production and development, and the code path is in the hot path of the hypervisor. The code to support VHE has been developed using ARM software models as ARMv8.1 hardware is not yet available, so it remains to be seen what performance gains will be achieved in practice with VHE.

7.10   FURTHER READING

Full-system virtualization of the ARM architecture is relatively new compared to x86. The earliest study of ARM hardware virtualization support was based on a software simulator and a simple hypervisor without SMP support, but due to the lack of hardware or a cycle-accurate simulator, no real performance evaluation was possible [173]. KVM/ARM was the first full-system virtualization ARM solution to use ARM hardware virtualization support and is described in further detail in [60]. In addition to KVM/ARM, a newer version of Xen targeting servers [188] has been developed using ARM hardware virtualization support. Because Xen is a bare metal hypervisor that does not leverage kernel functionality, it can be architected to run entirely in EL2 rather than using split-mode virtualization. At the same time, this requires a substantial commercial engineering effort. Porting Xen from x86 to ARM is difficult in part because all ARM-related code must be written from scratch. Even after getting Xen to work on one ARM platform, it must be manually ported to each different ARM device that Xen wants to support.

Various other approaches have considered different hypervisor structures. Microkernel approaches for hypervisors [75, 161] have been used to reduce the hypervisor TCB and run other hypervisor services in user mode. These approaches differ both in design and rationale from split-mode virtualization, which splits hypervisor functionality across privileged modes to leverage virtualization hardware support.

CHAPTER 8

Comparing ARM and x86 Virtualization Performance

This chapter presents a comparison of ARM and x86 virtualization performance on multicore server hardware, including measurements of two popular open-source ARM and x86 hypervisors, KVM and Xen. These hypervisors are useful to compare given their popularity and their different design choices. This work is based on a measurement study published in 2016 using state-of-the-art hardware for that time [58]. §8.1 first provides background regarding the current design of KVM and Xen. §8.2 describes the experimental setup used to measure ARM and x86 virtualization performance. §8.3 presents measurements of hypervisor performance based on running microbenchmarks to analyze low-level behavior. §8.4 presents measurements of hypervisor performance based on running application workloads to quantify performance for real applications running in VMs. Finally, like all chapters, we close with pointers for further reading.

8.1 KVM AND XEN OVERVIEW

To measure ARM and x86 virtualization performance, we used the widely used KVM and Xen hypervisor implementations to provide a realistic measure of performance in real software. We present a brief review of KVM and Xen based on their current implementations so the reader can more easily understand the subsequent performance results. Both KVM and Xen now make use of hardware virtualization support on both ARM and x86. Recall that Xen is a type-1 hypervisor that is a separate standalone software component which runs directly on the hardware and provides a virtual machine abstraction to VMs running on top of the hypervisor. KVM is a type-2 hypervisor that runs an existing operating system on the hardware and run both VMs and applications on top of the OS.

An advantage of type-2 hypervisors over type-1 hypervisors is the reuse of existing OS code, specifically device drivers for a wide range of available hardware. Traditionally, a type-1 hypervisor suffers from having to reimplement device drivers for all supported hardware. However, Xen avoids this by only implementing a minimal amount of hardware support directly in the hypervisor and running a special privileged VM, dom0, which runs an existing OS such as Linux, leveraging all the existing device drivers for that OS. Xen then arbitrates I/O between normal VMs and dom0 such that dom0 can perform I/O using existing device drivers on behalf of other VMs.

Transitions from a VM to the hypervisor occur whenever the hypervisor exercises system control, such as processing interrupts or I/O. The hypervisor transitions back to the VM once it has completed its work managing the hardware, letting workloads in VMs continue executing. The cost of such transitions is pure overhead and can add significant latency in communication between the hypervisor and the VM. A primary goal in designing both hypervisor software and hardware support for virtualization is to reduce the frequency and cost of transitions as much as possible.

As discussed in Chapter 6, VMs can run guest OSes with standard device drivers for I/O, but because they do not have direct access to hardware, the hypervisor would need to emulate real I/O devices in software resulting in frequent transitions between the VM and the hypervisor, making each interaction with the emulated device an order of magnitude slower than communicating with real hardware. Alternatively, direct passthrough of I/O from a VM to the real I/O devices can be done using device assignment, but this requires more expensive hardware support and complicates VM migration. Instead, the most common approach is paravirtual I/O in which custom device drivers are used in VMs for virtual devices supported by the hypervisor, and the interface between the VM device driver and the virtual device is specifically designed to optimize interactions between the VM and the hypervisor and facilitate fast I/O. KVM uses an implementation of the Virtio [154] protocol for disk and networking support, and Xen uses its own implementation referred to simply as **Xen PV**. In KVM, the virtual device backend is implemented in the host OS, while in Xen it is implemented in the dom0 kernel. A key potential performance advantage for KVM is that the virtual device implementation in the KVM host kernel has full access to all of the machine's hardware resources, including VM memory, while the Xen virtual device implementation lives in a separate VM, dom0, which only has access to memory and hardware resources specifically allocated to it by the Xen hypervisor. On the other hand, Xen provides a stronger isolation between the virtual device implementation and the VM.

The differences between Xen and KVM affect how they use hardware virtualization support on x86 vs. ARM. On x86, running Linux in root mode does not require any changes to Linux, so both KVM and Xen map equally well to the x86 architecture by running the hypervisor in root mode. Root mode does not limit nor change how CPU privilege levels are used. On ARM, while KVM uses split-mode virtualization across EL1 and EL2 as discussed in Chapter 7, Xen as a type-1 hypervisor design maps easily to the ARM architecture, running the entire hypervisor in EL2 and running VM userspace and VM kernel in EL0 and EL1, respectively. KVM only runs the minimal set of hypervisor functionality in EL2 to be able to switch between VMs and the host and emulates all virtual devices in the host OS running in EL1 and EL0. When a KVM VM performs I/O it involves trapping to EL2, switching to host EL1, and handling the I/O request in the host. Xen only emulates the GIC in EL2 and offloads all other I/O handling to dom0, which, like any other VM, runs its kernel in EL1. When a Xen VM performs I/O, it involves trapping to the hypervisor, signaling dom0, scheduling dom0, and handling the I/O request in dom0.

8.2   EXPERIMENTAL DESIGN

To evaluate the performance of ARM and x86 virtualization, both microbenchmarks and real application workloads were run on server hardware available as part of the Utah CloudLab [52]. ARMv8.1 hardware is not yet available at the time of this writing, so ARMv8 hardware was used. ARM measurements were done using HP Moonshot m400 servers, each equipped with a 64-bit ARMv8-A 2.4 GHz Applied Micro Atlas SoC with 8 physical CPU cores. Each m400 node in the cloud infrastructure is equipped with 64 GB of RAM, a 120 GB SATA3 SSD for storage, and a Dual-port Mellanox ConnectX-3 10GbE NIC. x86 measurements were done using Dell PowerEdge r320 servers, each equipped with a 64-bit Xeon 2.1 GHz ES-2450 with 8 physical CPU cores. Hyperthreading was disabled on the r320 nodes to provide a similar hardware configuration to the ARM servers. Each r320 node in the cloud infrastructure is equipped with 16 GB of RAM, a 4x500 GB 7200 RPM SATA RAID5 HD for storage, and a Dual-port Mellanox MX354A 10GbE NIC. All servers are connected via 10 GbE, and the interconnecting network switch [87] easily handles multiple sets of nodes communicating with full 10 Gb bandwidth such that experiments involving networking between two nodes can be considered isolated and unaffected by other traffic in the system. Using 10 Gb Ethernet was important, as many benchmarks were unaffected by virtualization when run over 1 Gb Ethernet, because the network itself became the bottleneck.

To provide comparable measurements, software environments across all hardware platforms and all hypervisors were kept the same as much as possible. KVM in Linux 4.0-rc4 with QEMU 2.2.0 and Xen 4.5.0 were used, which were the most recent stable versions of these hypervisors available at the time of the measurements. KVM was configured with its standard VHOST networking feature, allowing data handling to occur in the kernel instead of userspace, and with `cache=none` for its block storage devices. This is a commonly used I/O configuration for real KVM deployments that allows the guest access to the disk write cache, but disallows further memory caching that would sacrifice data persistence for additional performance gains. Xen was configured with its in-kernel block and network backend drivers to provide best performance and reflect the most commonly used I/O configuration for Xen deployments. All hosts and VMs used Ubuntu 14.04 [121] with the same Linux 4.0-rc4 kernel and software configuration for all machines. A few patches were applied to support the various hardware configurations, such as adding support for the APM X-Gene PCI bus for the HP m400 servers. All VMs used paravirtualized I/O, typical of cloud infrastructure deployments such as Amazon EC2, instead of device passthrough, due to the absence of an IOMMU in the test environment used.

Benchmarks were run both natively on the hosts and in VMs. Each physical or virtual machine instance used for running benchmarks was configured as a 4-way SMP with 12 GB of RAM to provide a common basis for comparison. This involved three configurations: (1) running natively on Linux capped at 4 cores and 12 GB RAM; (2) running in a VM using KVM with 8 cores and 16 GB RAM with the VM capped at 4 virtual CPUs (VCPUs) and 12 GB RAM; and (3) running in a VM using Xen with dom0, the privileged domain used by Xen with direct hardware access, capped at 4 cores and 4 GB RAM and the VM capped at 4 VCPUs and 12 GB RAM. Because KVM configures the total hardware available while Xen configures the hardware dedicated to dom0, the configuration parameters are different. The effect nevertheless is the same, which is to leave the hypervisor with 4 cores and 4 GB RAM to use outside of what is used by the VM. Multicore configurations were used to reflect real-world server deployments. The memory limit was used to ensure a fair comparison across all hardware configurations given the RAM available on the x86 servers and the need to also provide RAM for use by the hypervisor when running VMs. For benchmarks that involve clients interfacing with the server, the clients were run natively on Linux and configured to use the full hardware available.

To improve measurement precision and mimic recommended configuration best practices [187], each VCPU was pinned to a specific physical CPU (PCPU) and generally no other work was scheduled on that PCPU. In KVM, all of the host's device interrupts and processes were assigned to run on a specific set of PCPUs and each VCPU was pinned to a dedicated PCPU from a separate set of PCPUs. In Xen, dom0 was configured to run on a set of PCPUs while domU was run on a separate set of PCPUs. Each VCPU of both dom0 and domU was pinned to its own PCPU.

8.3   MICROBENCHMARK RESULTS

We present measurements for seven microbenchmarks that quantify various low-level aspects of hypervisor performance, as listed in Table 8.1. Some of these measurements are similar to those presented in Chapter 7, but are done with commercial 64-bit server hardware instead of development boards, providing a more realistic comparison of server virtualization performance. A primary performance cost in running in a VM is how much time must be spent outside the VM, which is time not spent running the workload in the VM and therefore is virtualization overhead compared to native execution. Therefore, the microbenchmarks are designed to measure time spent handling a trap from the VM to the hypervisor, including time spent on transitioning between the VM and the hypervisor, time spent processing interrupts, time spent switching between VMs, and latency added to I/O. Table 8.2 presents the results from running these microbenchmarks on both ARM and x86 server hardware. Measurements are shown in cycles instead of time to provide a useful comparison across server hardware with different CPU frequencies. Measurements were obtained using cycle counters and ARM hardware timer counters to ensure consistency across multiple CPUs, and instruction barriers were used before and after taking timestamps to avoid out-of-order execution or pipelining from skewing our measurements.

The Hypercall microbenchmark shows that transitioning from a VM to the hypervisor on ARM can be significantly faster than x86, as shown by the Xen ARM measurement, which takes less than a third of the cycles that Xen or KVM on x86 take. The ARM architecture provides a separate CPU mode with its own register bank to run an isolated type-1 hypervisor like Xen. Transitioning from a VM to a type-1 hypervisor requires little more than context switching the general purpose registers as running the two separate execution contexts, VM and the hypervisor,

Table 8.1: Microbenchmarks

Name	Description
Hypercall	Transition from VM to hypervisor and return to VM without doing any work in the hypervisor. Measures bidirectional base transition cost of hypervisor operations.
Interrupt Controller Trap	Trap from VM to emulated interrupt controller then return to VM. Measures a frequent operation for many device drivers and baseline for accessing I/O devices emulated in the hypervisor.
Virtual IPI	Issue a virtual IPI from a VCPU to another VCPU running on a different PCPU, both PCPUs executing VM code. Measures time between sending the virtual IPI until the receiving VCPU handles it, a frequent operation in multicore OSes that affects many workloads.
Virtual IRQ Completion	VM acknowledging and completing a virtual interrupt. Measures a frequent operation that happens for every injected virtual interrupt.
VM Switch	Switching from one VM to another on the same physical core. Measures a central cost when oversubscribing physical CPUs.
I/O Latency Out	Measures latency between a driver in the VM signaling the virtual I/O device in the hypervisor and the virtual I/O device receiving the signal. For KVM, this involves trapping to the host kernel. For Xen, this involves trapping to Xen then raising a virtual interrupt to Dom0.
I/O Latency In	Measures latency between the virtual I/O device in the hypervisor signaling the VM and the VM receiving the corresponding virtual interrupt. For KVM, this involves signaling the VCPU thread and injecting a virtual interrupt for the Virtio device. For Xen, this involves trapping to Xen then raising a virtual interrupt to DomU.

is supported by the separate ARM hardware state for EL2. While ARM implements additional register state to support the different execution context of the hypervisor, x86 transitions from a VM to the hypervisor by switching from non-root to root mode which requires context switching the entire CPU register state to the VMCS in memory, which is much more expensive even with hardware support.

However, the Hypercall microbenchmark also shows that transitioning from a VM to the hypervisor on ARM is more than an order of magnitude more expensive for type-2 hypervisors

Table 8.2: Microbenchmark measurements (cycle counts)

Microbenchmark	ARM		x86	
	KVM	Xen	KVM	Xen
Hypercall	6,500	376	1,300	1,228
Interrupt Controller Trap	7,370	1,356	2,384	1,734
Virtual IPI	11,557	5,978	5,230	5,562
Virtual IRQ Completion	71	71	1,556	1,464
VM Switch	10,387	8,799	4,812	10,534
I/O Latency Out	6,024	16,491	560	11,262
I/O Latency In	13,872	15,650	18,923	10,050

like KVM than for type-1 hypervisors like Xen. This is because although all VM traps are handled in EL2, a type-2 hypervisor is integrated with a host kernel and both run in EL1. This results in four additional sources of overhead. First, transitioning from the VM to the hypervisor involves not only trapping to EL2, but also returning to the host OS in EL1, incurring a double trap cost. Second, because the host OS and the VM both run in EL1 and ARM hardware does not provide any features to distinguish between the host OS running in EL1 and the VM running in EL1, software running in EL2 must context switch all the EL1 system register state between the VM guest OS and the type-2 hypervisor host OS, incurring added cost of saving and restoring EL1 register state. Third, because the host OS runs in EL1 and needs full access to the hardware, the hypervisor must disable traps to EL2 and Stage-2 translation from EL2 while switching from the VM to the hypervisor, and enable them when switching back to the VM again. Fourth, because the type-2 hypervisor runs in EL1 but needs to access VM control register state such as the VGIC state, which can only be accessed from EL2, there is additional overhead to read and write the VM control register state in EL2. The type-2 hypervisor can either jump back and forth between EL1 and EL2 to access the control register state when needed, or it can copy the full register state to memory while it is still in EL2, return to the host OS in EL1 and read and write the memory copy of the VM control state, and then finally copy the state from memory back to the EL2 control registers when the hypervisor is running in EL2 again. Both methods incur much overhead, but jumping back and forward between EL1 and EL2 makes the software implementation complicated and difficult to maintain. Therefore, the KVM/ARM implementation currently takes the second approach of reading and writing all VM control registers in EL2 during each transition between the VM and the hypervisor.

While the cost of the trap between CPU modes itself is not very high as shown in Chapter 7, these measurements show that there is a substantial cost associated with saving and restoring register state to switch between EL2 and the host in EL1. Table 8.3 provides a breakdown of the cost of context switching the relevant register state when performing the Hypercall microbenchmark

Table 8.3: KVM/ARM Hypercall analysis (cycle counts)

Register State	Save	Restore
GP Regs	152	184
FP Regs	282	310
EL1 System Regs	230	511
VGIC Regs	3,250	181
Timer Regs	104	106
EL2 Config Regs	92	107
EL2 Virtual Memory Regs	92	107

measurement on KVM/ARM. Context switching consists of saving register state to memory and restoring the new context's state from memory to registers. The cost of saving and restoring this state accounts for almost all of the Hypercall time, indicating that context switching state is the primary cost due to KVM/ARM's design, not the cost of extra traps. Unlike Xen ARM which only incurs the relatively small cost of saving and restoring the general-purpose (GP) registers, KVM/ARM saves and restores much more register state at much higher cost. Note that for ARM, the overall cost of saving register state, when transitioning from a VM to the hypervisor, is much more expensive than restoring it, when returning back to the VM from the hypervisor, due to the cost of reading the VGIC register state.

Unlike on ARM, both x86 hypervisors spend a similar amount of time transitioning from the VM to the hypervisor. Since both KVM and Xen leverage the same x86 hardware mechanism for transitioning between VM and hypervisor, they have similar performance. Both x86 hypervisors run in root mode and run their VMs in non-root mode, and switching between the two modes involves switching a substantial portion of the CPU register state to the VMCS in memory. Switching this state to memory is fast on x86, because it is performed by hardware in the context of a trap or as a result of executing a single instruction. In contrast, ARM provides a separate CPU mode for the hypervisor with separate registers, and ARM only needs to switch state to memory when running a different execution context in EL1. ARM can be much faster, as in the case of Xen ARM which does its hypervisor work in EL2 and does not need to context switch much register state, or it can be much slower, as in the case of KVM/ARM which context switches more register state without the benefit of hardware support like x86.

The large difference between Xen ARM and KVM/ARM in the cost of transitioning between the VM and hypervisor, as shown by the Hypercall measurement, results in Xen ARM being significantly faster at handling interrupt related traps, because Xen ARM emulates the ARM GIC interrupt controller directly in the hypervisor running in EL2. In contrast, KVM/ARM emulates the GIC in the part of the hypervisor running in EL1. Therefore, operations such as accessing registers in the emulated GIC, sending virtual IPIs, and receiving virtual interrupts are much faster on Xen ARM than KVM/ARM. This is shown in Table 8.2 in the measurements for the Interrupt Controller trap and Virtual IPI microbenchmarks, in which Xen ARM is faster than KVM/ARM by roughly the same difference as for the Hypercall microbenchmark.

However, Table 8.2 shows that for the remaining microbenchmarks, Xen ARM does not enjoy a large performance advantage over KVM/ARM and in fact does worse for some of the microbenchmarks. The reasons for this differ from one microbenchmark to another: For the Virtual IRQ Completion microbenchmark, both KVM/ARM and Xen ARM are very fast because the ARM hardware includes support for completing interrupts directly in the VM without trapping to the hypervisor. The microbenchmark runs much faster on ARM than x86 because the latter has to trap to the hypervisor. More recently, vAPIC support has been added to x86 with similar functionality to avoid the need to trap to the hypervisor so that newer x86 hardware with vAPIC support should perform more comparably to ARM [104].

For the VM Switch microbenchmark, Xen ARM is only slightly faster than KVM/ARM because both hypervisor implementations have to context switch the state between the VM being switched out and the one being switched in. Unlike the Hypercall microbenchmark where only KVM/ARM needed to context switch EL1 state and per VM EL2 state, in this case both KVM and Xen ARM need to do this, and Xen ARM therefore does not directly benefit from its faster VM-to-hypervisor transition. Xen ARM is still slightly faster than KVM, however, because to switch between VMs, Xen ARM simply traps to EL2 and performs a single context switch of the EL1 state, while KVM/ARM must switch the EL1 state from the VM to the host OS and then again from the host OS to the new VM. Finally, KVM/ARM also has to disable and enable traps and Stage-2 translation on each transition, which Xen ARM does not have to do.

For the I/O Latency microbenchmarks, a surprising result is that Xen ARM is slower than KVM/ARM in both directions. These microbenchmarks measure the time from when a network I/O event is initiated by a sender until the receiver is notified, not including additional time spent transferring data. I/O latency is an especially important metric for real-time sensitive operations and many networking applications. The key insight to understanding the results is to see that Xen ARM does not benefit from its faster VM-to-hypervisor transition mechanism in this case because Xen ARM must switch between two separate VMs, dom0 and a domU, to process network I/O. Type-1 hypervisors only implement a limited set of functionality in the hypervisor directly, namely scheduling, memory management, the interrupt controller, and timers for Xen ARM. All other functionality, for example network and storage drivers are implemented in the special privileged VM, dom0. Therefore, a VM performing I/O has to communicate with dom0 and not just the Xen hypervisor, which means not just trapping to EL2, but also going to EL1 to run dom0.

I/O Latency Out is much worse on Xen ARM than KVM/ARM. When KVM/ARM sends a network packet, it traps to the hypervisor, which involves context switching the EL1 state, and then the host OS instance directly sends the data on the physical network. Xen ARM, on the other hand, must trap from the VM to the hypervisor, which then signals a different VM, dom0, and dom0 then sends the data on the physical network. This signaling between VMs on Xen is slow for two main reasons. First, because the VM and dom0 run on different physical CPUs, Xen must send a physical IPI from the CPU running the VM to the CPU running dom0. Second, Xen actually switches from dom0 to a special VM, called the idle domain, when dom0 is idling and waiting for I/O. Thus, when Xen signals dom0 to perform I/O on behalf of a VM, it must perform a VM switch from the idle domain to dom0. Changing the configuration of Xen to pinning both the VM and dom0 to the same physical CPU or not specifying any pinning at all resulted in similar or worse results than reported in Table 8.2, so the qualitative results are not specific to this configuration.

It is interesting to note that KVM x86 is much faster than everything else on I/O Latency Out. KVM on both ARM and x86 involve the same control path of transitioning from the VM to the hypervisor. While the path is conceptually similar to half of the path for the Hypercall microbenchmark, the result for the I/O Latency Out microbenchmark is not 50% of the Hypercall cost on either platform. The reason is that for KVM x86, transitioning from the VM to the hypervisor accounts for around only 40% of the Hypercall cost, and transitioning from the hypervisor to the VM accounts for most of the rest (a few cycles are spent handling the noop hypercall in the hypervisor). On ARM, it is much more expensive to transition from the VM to the hypervisor than from the hypervisor to the VM, because reading back the VGIC state is expensive, as shown in Table 8.3.

I/O Latency In behaves more similarly between Xen ARM and KVM/ARM, because both hypervisors perform similar low-level operations. Xen traps from dom0 running in EL1 to the hypervisor running in EL2 and signals the receiving VM, the reverse of the procedure described above, thereby sending a physical IPI and switching from the idle domain to the receiving VM in EL1. For KVM/ARM, the Linux host OS receives the network packet and wakes up the idle VM's VCPU thread and signals the CPU running the VCPU thread, thereby sending a physical IPI, and the VCPU thread then traps to EL2, switches the EL1 state from the host OS to the VM, and switches to the VM in EL1. The end result is that the cost is similar across both hypervisors, with KVM being slightly faster. While KVM/ARM is slower on I/O Latency In than I/O Latency Out because it performs more work on the incoming path, Xen has similar performance on both Latency I/O In and Latency I/O Out because it performs comparable low-level operations for both microbenchmarks.

8.4 APPLICATION BENCHMARK RESULTS

We next present measurements of a number of real application benchmark workloads to quantify how well the ARM virtualization extensions support different hypervisor software designs in the context of more realistic workloads. Table 8.4 lists the application workloads used, which include a mix of widely-used CPU and I/O intensive benchmark workloads. For workloads involving a client and a server, the client ran on a dedicated machine and the server ran on the configuration being measured, ensuring that the client was never saturated during any of our experiments. These

Table 8.4: Application benchmarks

Benchmark	Description
Kernbench	Kernel compilation by compiling the Linux 3.17.0 kernel using the allnoconfig for ARM using GCC 4.8.2.
Hackbench	`hackbench` [132] using unix domain sockets and 100 process groups running with 500 loops.
SPECjvm2008	`SPECjvm2008` [160] 2008 benchmark running several real life applications and benchmarks specifically chosen to benchmark the performance of the Java Runtime Environment. 15.02 release of the Linaro AArch64 port of OpenJDK was used run the benchmark.
Netperf	`netperf` v2.6.0 starting netserver on the server and running with its default parameters on the client in three modes: TCP_STREAM, TCP_MAERTS, and TCP_RR, measuring throughput transferring data from client to server, throughput transferring data from server to client, and latency, respectively.
Apache	`Apache` v2.4.7 Web server running `ApacheBench` v2.3 on the remote/local client, which measures the number of handled requests per second serving the index file of the GCC 4.4 manual using 100 concurrent requests.
Memcached	`memcached` v1.4.14 using the `memtier` benchmark v1.2.3 with its default parameters.
MySql	`MySQL` v14.14 (distrib 5.5.41) running the `SysBench` v.0.4.12 OLTP benchmark using the default configuration with 200 parallel transactions.

workloads were run natively and on both KVM and Xen on both ARM and x86, the latter to provide a baseline comparison. Table 8.5 shows the raw results for executing the workloads on both ARM and x86 servers.

Given the differences in hardware platforms, we focus not on measuring absolute performance, but rather the relative performance differences between virtualized and native execution on each platform. Figure 8.1 shows the performance overhead of KVM and Xen on ARM and x86 compared to native execution on the respective platform. All numbers are normalized to 1 for native performance, so that lower numbers represent better performance. Unfortunately, the Apache benchmark could not run on Xen x86 because it caused a kernel panic in dom0. This problem was reported to the Xen developer community and it apparently was due to a Mellanox network driver bug exposed by Xen's I/O model.

Figure 8.1 shows that the application performance on KVM and Xen on ARM and x86 does not appear well correlated with their respective performance on the microbenchmarks shown in

Table 8.5: Application benchmark raw performance

Name	CPU	Native	KVM	Xen
Kernbench (s)	ARM	49.11	50.49	49.83
	x86	28.91	27.12	27.56
Hackbench (s)	ARM	15.65	17.38	16.55
	x86	6.04	6.66	6.57
SPECjvm2008 (ops/min)	ARM	62.43	61.69	61.91
	x86	140.76	140.64	141.80
TCP_RR (trans/s)	ARM	23,911	11.591	10,253
	x86	21,089	11,490	7,661
TCP_STREAM (Mb/s)	ARM	5,924	5,603	1,662
	x86	9,174	9,287	2,353
TCP_MAERTS (Mb/s)	ARM	6,051	6,059	3,778
	x86	9,148	8,817	5,948
Apache (trans/s)	ARM	6,526	4,846	3,539
	x86	10,585	9,170	N/A
Memcached (ops/s)	ARM	110,865	87,811	84,118
	x86	263,302	170,359	226,403
MySQL (s)	ARM	13.72	15.76	15.02
	x86	7.21	9.08	8.75

Table 8.2. Xen ARM has by far the lowest VM to hypervisor transition costs and the best performance for most of the microbenchmarks, yet its performance lags behind KVM/ARM on many of the application benchmarks. KVM ARM substantially outperforms Xen ARM on the various netperf benchmarks, TCP_STREAM, TCP_MAERTS, and TCP_RR, as well as Apache and Memcached, and performs only slightly worse on the rest of the application benchmarks. Xen ARM also does generally worse than KVM x86. Clearly, the differences in microbenchmark performance do not result in the same differences in real application performance.

Xen ARM achieves its biggest performance gain vs. KVM/ARM on Hackbench. Hackbench involves running lots of threads that are sleeping and waking up, requiring frequent IPIs for rescheduling. Xen ARM performs virtual IPIs much faster than KVM/ARM, roughly a factor of two. Despite this microbenchmark performance advantage on a workload performing frequent virtual IPIs, the resulting difference in Hackbench performance overhead is small, only 5% of native performance. Overall, across CPU-intensive workloads such as Kernbench, Hackbench, and SPECjvm, the performance differences among the different hypervisors across different architectures is small.

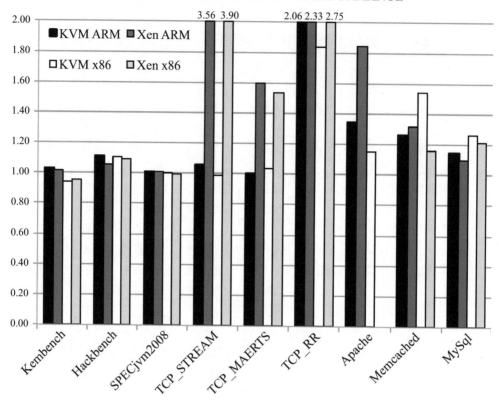

Figure 8.1: Application benchmark performance normalized to native execution.

Figure 8.1 shows that the largest differences in performance are for the I/O-intensive workloads. We first take a closer look at the netperf results. Netperf TCP_RR is an I/O latency benchmark, which sends a 1 byte packet from a client to the Netperf server running in the VM, and the Netperf server sends the packet back to the client, and the process is repeated for 10 s. For the netperf TCP_RR benchmark, both hypervisors show high overhead compared to native performance, but Xen is noticeably worse than KVM. Taking a closer look at KVM/ARM vs. Xen ARM, there are two main reasons why Xen performs worse. First, Xen's I/O latency is higher than KVM's as measured and explained by the I/O Latency In and Out microbenchmarks in §8.3. Second, Xen does not support zero-copy I/O, but must map a shared page between dom0 and the VM using Xen's grant tables, a generic mechanism for memory sharing between domains. Xen must copy data between the memory buffer used for DMA in dom0 and the granted memory buffer from the VM. Each data copy incurs more than 3 μs of additional latency because of the complexities of establishing and utilizing the shared page via the grant table mechanism across VMs, even though only a single byte of data needs to be copied.

Although Xen ARM can transition between the VM and hypervisor more quickly than KVM, Xen cannot utilize this advantage for the TCP_RR workload, because Xen must engage dom0 to perform I/O on behalf of the VM, which results in several VM switches between idle domains and dom0 or domU, and because Xen must perform expensive page mapping operations to copy data between the VM and dom0. This is a direct consequence of Xen's software architecture and I/O model based on domains and a strict I/O isolation policy. Xen ends up spending so much time communicating between the VM and dom0 that it completely dwarfs its low Hypercall cost for the TCP_RR workload and ends up having more overhead than KVM/ARM, due to Xen's software architecture and I/O model in particular.

The hypervisor software architecture is also a dominant factor in other aspects of the netperf results. For the netperf TCP_STREAM benchmark, KVM has almost no overhead for x86 and ARM while Xen has more than 250% overhead. The reason for this large difference in performance is again due to Xen's lack of zero-copy I/O support, in this case particularly on the network receive path. The netperf TCP_STREAM benchmark sends large quantities of data from a client to the netperf server in the VM. Xen's dom0, running Linux with the physical network device driver, cannot configure the network device to DMA the data directly into guest buffers, because dom0 does not have access to the VM's memory. When Xen receives data, it must configure the network device to DMA the data into a dom0 kernel memory buffer, signal the VM for incoming data, let Xen configure a shared memory buffer, and finally copy the incoming data from the dom0 kernel buffer into the virtual device's shared buffer. KVM, on the other hand, has full access to the VM's memory and maintains shared memory buffers in the Virtio rings [154], such that the network device can DMA the data directly into a guest-visible buffer, resulting in significantly less overhead.

Furthermore, previous work [156] and discussions with the Xen maintainers confirm that supporting zero copy on x86 is problematic for Xen given its I/O model because doing so requires signaling all physical CPUs to locally invalidate TLBs when removing grant table entries for shared pages, which proved more expensive than simply copying the data [92]. As a result, previous efforts to support zero copy on Xen x86 were abandoned. Xen ARM lacks the same zero copy support because the dom0 network backend driver uses the same code as on x86. Whether zero copy support for Xen can be implemented efficiently on ARM, which has hardware support for broadcasted TLB invalidate requests across multiple PCPUs, remains to be investigated.

For the netperf TCP_MAERTS benchmark, Xen also has substantially higher overhead than KVM. The benchmark measures the network transmit path from the VM, the converse of the TCP_STREAM benchmark which measured the network receive path to the VM. It turns out that the Xen performance problem is due to a regression in Linux introduced in Linux v4.0-rc1 in an attempt to fight bufferbloat, and has not yet been fixed beyond manually tuning the Linux TCP configuration in the guest OS [123]. Using an earlier version of Linux or tuning the TCP configuration in the guest using sysfs significantly reduced the overhead of Xen on the TCP_MAERTS benchmark.

Other than the netperf workloads, the application workloads with the highest overhead were Apache and Memcached. The performance bottleneck for KVM and Xen on ARM was due to network interrupt processing and delivery of virtual interrupts. Delivery of virtual interrupts is more expensive than handling physical IRQs on bare-metal, because it requires switching from the VM to the hypervisor and injecting a virtual interrupt to the VM and switching back to the VM. Additionally, Xen and KVM both handle all virtual interrupts using a single VCPU, which, combined with the additional virtual interrupt delivery cost, fully utilizes the underlying PCPU. If changes are made so that virtual interrupts are distributed across multiple VCPUs, KVM performance overhead dropped from 35% to 14% on Apache and from 26% to 8% on Memcached, while the Xen performance overhead dropped from 84% to 16% on Apache and from 32% to 9% on Memcached.

In summary, while the VM-to-hypervisor transition cost for a type-1 hypervisor like Xen is much lower on ARM than for a type-2 hypervisor like KVM, this difference is not easily observed for the application workloads. The reason is that type-1 hypervisors typically only support CPU, memory, and interrupt virtualization directly in the hypervisors. CPU and memory virtualization has been highly optimized directly in hardware and, ignoring one-time page fault costs at startup, is performed largely without the hypervisor's involvement. That leaves only interrupt virtualization, which is indeed much faster for type-1 hypervisor on ARM, confirmed by the Interrupt Controller Trap and Virtual IPI microbenchmarks shown in §8.3. While this contributes to Xen's slightly better Hackbench performance, the resulting application performance benefit overall is modest.

However, when VMs perform I/O operations such as sending or receiving network data, type-1 hypervisors like Xen typically offload such handling to separate VMs to avoid having to re-implement all device drivers for the supported hardware and to avoid running a full driver and emulation stack directly in the type-1 hypervisor, which would significantly increase the Trusted Computing Base and increase the attack surface of the hypervisor. Switching to a different VM to perform I/O on behalf of the VM has very similar costs on ARM compared to a type-2 hypervisor approach of switching to the host on KVM. Additionally, KVM on ARM benefits from the hypervisor having privileged access to all physical resources, including the VM's memory, and from being directly integrated with the host OS, allowing for optimized physical interrupt handling, scheduling, and processing paths in some situations.

Despite the inability of both KVM and Xen to leverage the potential fast path of trapping from a VM running in EL1 to the hypervisor in EL2 without the need to run additional hypervisor functionality in EL1, our measurements show that both KVM and Xen on ARM can provide virtualization overhead similar to, and in some cases better than, their respective x86 counterparts. Furthermore, as discussed in Chapter 7, the introduction of ARM VHE could make it possible for KVM to better leverage the fast path of trapping to EL2 only by running all of KVM in EL2, resulting in further performance improvements for KVM on ARM.

8.5   FURTHER READING

Much work has been done on analyzing the performance of x86 virtualization. For example, Heo and Taheri [86] analyze the performance of various latency-sensitive workloads on VMware vSphere on x86, Buell et al. [41] investigate the performance of both complex computation-intensive as well as latency-sensitive workloads on VMware vSphere on multicore systems, and Bhargava et al. [35] finds that two-level TLB misses can be very expensive for any hypervisor using either Intel's or AMD's hardware support for virtualization.

Relatively fewer studies have been done on the performance of ARM virtualization, with most focused on ARMv7 development hardware [60, 79, 133, 149]. Since ARM virtualization support is more recent, little work has been done comparing x86 and ARM virtualization performance. The work reported here is based on the first in-depth measurement study of x86 vs. ARM virtualization [58].

Bibliography

[1] Darren Abramson, Jeff Jackson, Sridhar Muthrasanallur, Gil Neiger, Greg Regnier, Rajesh Sankaran, Ioannis Schoinas, Rich Uhlig, Balaji Vembu, and John Wiegert. Intel virtualization technology for directed I/O. *Intel Technology Journal*, 10(3):179–192, 2006. 79

[2] ACM SIGOPS. SIGOPS hall of fame award. http://www.sigops.org/award-hof.html 29, 51

[3] Keith Adams and Ole Agesen. A comparison of software and hardware techniques for x86 virtualization. In *Proc. of the 12th International Conference on Architectural Support for Programming Languages and Operating Systems (ASPLOS-XII)*, pages 2–13, 2006. DOI: 10.1145/1168857.1168860 43, 55, 61, 69

[4] Ole Agesen, Alex Garthwaite, Jeffrey Sheldon, and Pratap Subrahmanyam. The evolution of an x86 virtual machine monitor. *Operating Systems Review*, 44(4):3–18, 2010. DOI: 10.1145/1899928.1899930 51, 77

[5] Ole Agesen, Jim Mattson, Radu Rugina, and Jeffrey Sheldon. Software techniques for avoiding hardware virtualization exits. In *Proc. of the USENIX Annual Technical Conference (ATC)*, pages 373–385, 2012. https://www.usenix.org/conference/atc12/technical-sessions/presentation/agesen 68, 75, 77

[6] Ole Agesen and Jeffrey Sheldon. Personal communication (VMware), 2015. 68

[7] Irfan Ahmad, Jennifer M. Anderson, Anne M. Holler, Rajit Kambo, and Vikram Makhija. An analysis of disk performance in vmware esx server virtual machines. In *IEEE International Workshop on Workload Characterization*, pages 65–76, 2003. 46, 51

[8] Jeongseob Ahn, Seongwook Jin, and Jaehyuk Huh. Revisiting hardware-assisted page walks for virtualized systems. In *Proc. of the 39th International Symposium on Computer Architecture (ISCA)*, pages 476–487, 2012. DOI: 10.1109/ISCA.2012.6237041 77

[9] Brian Aker. Memslap—load testing and benchmarking a server. http://docs.libmemcached.org/bin/memslap.html Accessed: August 2016. 113

[10] Altera Corporation. Arria 10 Avalon-ST interface with SR-IOV PCIe solutions: User guide. https://www.altera.com/en_US/pdfs/literature/ug/ug_a10_pcie_sriov.pdf, May 2016. Accessed: August 2016. 109

[11] AMD Corporation. AMD I/O virtualization technology (IOMMU) specification. Revision 2.62. http://support.amd.com/TechDocs/48882_IOMMU.pdf, 2015. Accessed: August 2016. 102, 105

[12] Nadav Amit, Muli Ben-Yehuda, Dan Tsafrir, and Assaf Schuster. vIOMMU: Efficient IOMMU emulation. In *Proc. of the USENIX Annual Technical Conference (ATC)*, 2011. https://www.usenix.org/conference/usenixatc11/viommu-efficient-iommu-emulation xvii, 105, 121

[13] Nadav Amit, Abel Gordon, Nadav Har'El, Muli Ben-Yehuda, Alex Landau, Assaf Schuster, and Dan Tsafrir. Bare-metal performance for virtual machines with exitless interrupts. *Communications of the ACM*, 59(1):108–116, 2016. DOI: 10.1145/2845648 xvii, 113, 115

[14] Nadav Amit, Dan Tsafrir, and Assaf Schuster. VSwapper: A memory swapper for virtualized environments. In *Proc. of the 19th International Conference on Architectural Support for Programming Languages and Operating Systems (ASPLOS-XIX)*, pages 349–366, 2014. DOI: 10.1145/2541940.2541969 xvii

[15] Nadav Amit, Dan Tsafrir, Assaf Schuster, Ahmad Ayoub, and Eran Shlomo. Virtual CPU validation. In *Proc. of the 25th ACM Symposium on Operating Systems Principles (SOSP)*, pages 311–327, 2015. DOI: 10.1145/2815400.2815420 xvii, 66, 68

[16] Jeremy Andrus, Christoffer Dall, Alexander Van't Hof, Oren Laadan, and Jason Nieh. Cells: A virtual mobile smartphone architecture. In *Proc. of the 23rd ACM Symposium on Operating Systems Principles (SOSP)*, pages 173–187, 2011. DOI: 10.1145/2043556.2043574 xiv, xvi

[17] Apache HTTP Server Benchmarking Tool. https://httpd.apache.org/docs/2.2/programs/ab.html. Accessed: August 2016. 113

[18] The Apache HTTP Server Project. http://httpd.apache.org. Accessed: August 2016. 113

[19] Andrea Arcangeli, Izik Eidus, and Chris Wright. Increasing memory density by using KSM. In *Proc. of the 2009 Ottawa Linux Symposium (OLS)*, pages 19–28, 2009. 75

[20] ARM Ltd. ARM architecture reference manual (ARM DDI 0100I), 2005. 25

[21] ARM Ltd. ARM generic interrupt controller architecture version 2.0 ARM IHI 0048B, June 2011. 130

[22] ARM Ltd. ARM generic interrupt controller architecture version 3.0 and version 4.0 ARM IHI 0069C, July 2016. 131

[23] ARM Ltd. ARM system memory management unit architecture specification SMMU architecture version 2.0. `http://infocenter.arm.com/help/topic/com.arm.doc.ih i0062d.c/IHI0062D_c_system_mmu_architecture_specification.pdf`, 2016. Accessed: August 2016. 102

[24] Gaurav Banga, Sergei Vorobiev, Deepak Khajura, Ian Pratt, Vikram Kapoor, and Simon Crosby. Seamless management of untrusted data using virtual machines, September 29 2015. U.S. Patent 9,148,428. `https://www.google.com/patents/US9148428` xvi

[25] Ricardo A. Baratto, Leonard N. Kim, and Jason Nieh. THINC: A virtual display architecture for thin-client computing. In *Proc. of the 20th ACM Symposium on Operating Systems Principles (SOSP)*, pages 277–290, 2005. DOI: 10.1145/1095810.1095837 xvi

[26] Ricardo A. Baratto, Shaya Potter, Gong Su, and Jason Nieh. MobiDesk: Mobile virtual desktop computing. In *Proc. of the 10th Annual International Conference on Mobile Computing and Networking (MobiCom)*, pages 1–15, 2004. DOI: 10.1145/1023720.1023722 xvi

[27] Paul Barham, Boris Dragovic, Keir Fraser, Steven Hand, Timothy L. Harris, Alex Ho, Rolf Neugebauer, Ian Pratt, and Andrew Warfield. Xen and the art of virtualization. In *Proc. of the 19th ACM Symposium on Operating Systems Principles (SOSP)*, pages 164–177, 2003. DOI: 10.1145/945445.945462 xiv, 7, 43, 44, 45, 46

[28] Kenneth C. Barr, Prashanth P. Bungale, Stephen Deasy, Viktor Gyuris, Perry Hung, Craig Newell, Harvey Tuch, and Bruno Zoppis. The VMware mobile virtualization platform: Is that a hypervisor in your pocket? *Operating Systems Review*, 44(4):124–135, 2010. DOI: 10.1145/1899928.1899945 48, 51, 123

[29] Adam Belay, Andrea Bittau, Ali José Mashtizadeh, David Terei, David Mazières, and Christos Kozyrakis. Dune: Safe user-level access to privileged CPU features. In *Proc. of the 10th Symposium on Operating System Design and Implementation (OSDI)*, pages 335–348, 2012. `https://www.usenix.org/conference/osdi12/technical-sessions/p resentation/belay` xvi

[30] Adam Belay, George Prekas, Ana Klimovic, Samuel Grossman, Christos Kozyrakis, and Edouard Bugnion. IX: A protected dataplane operating system for high throughput and low latency. In *Proc. of the 11th Symposium on Operating System Design and Implementation (OSDI)*, pages 49–65, 2014. `https://www.usenix.org/conference/osdi14/techni cal-sessions/presentation/belay` xvi

[31] Adam Belay, George Prekas, Mia Primorac, Ana Klimovic, Samuel Grossman, Christos Kozyrakis, and Edouard Bugnion. The IX operating system: Combining low latency, high throughput, and efficiency in a protected dataplane. *ACM Transactions on Computer Systems*, 34(4):11:1–11:39, December 2016. DOI: 10.1145/2997641 xvi

[32] Fabrice Bellard. QEMU, a fast and portable dynamic translator. In *USENIX Annual Technical Conference, FREENIX Track*, pages 41–46, 2005. http://www.usenix.org/events/usenix05/tech/freenix/bellard.html 6, 62, 63

[33] Muli Ben-Yehuda, Michael D. Day, Zvi Dubitzky, Michael Factor, Nadav Har'El, Abel Gordon, Anthony Liguori, Orit Wasserman, and Ben-Ami Yassour. The turtles project: Design and implementation of nested virtualization. In *Proc. of the 9th Symposium on Operating System Design and Implementation (OSDI)*, pages 423–436, 2010. http://www.usenix.org/events/osdi10/tech/full_papers/Ben-Yehuda.pdf xvi, 22, 60, 72

[34] Muli Ben-Yehuda, Michael Factor, Eran Rom, Avishay Traeger, Eran Borovik, and Ben-Ami Yassour. Adding advanced storage controller functionality via low-overhead virtualization. In *Proc. of the 10th USENIX Conference on File and Storage Technologie (FAST)*, page 15, 2012. https://www.usenix.org/conference/fast12/adding-advanced-storage-controller-functionality-low-overhead-virtualization 121

[35] Ravi Bhargava, Ben Serebrin, Francesco Spadini, and Srilatha Manne. Accelerating two-dimensional page walks for virtualized systems. In *Proc. of the 13th International Conference on Architectural Support for Programming Languages and Operating Systems (ASPLOS-XIII)*, pages 26–35, 2008. DOI: 10.1145/1346281.1346286 71, 75, 77, 161

[36] Nathan L. Binkert, Bradford M. Beckmann, Gabriel Black, Steven K. Reinhardt, Ali G. Saidi, Arkaprava Basu, Joel Hestness, Derek Hower, Tushar Krishna, Somayeh Sardashti, Rathijit Sen, Korey Sewell, Muhammad Shoaib, Nilay Vaish, Mark D. Hill, and David A. Wood. The gem5 simulator. *SIGARCH Computer Architecture News*, 39(2):1–7, 2011. DOI: 10.1145/2024716.2024718 5

[37] Paolo Bonzini. The security state of KVM. https://lwn.net/Articles/619332/, 2014. 69

[38] David Brash. ARMv8-A architecture—2016 additions. https://community.arm.com/groups/processors/blog/2016/10/27/armv8-a-architecture-2016-additions 127

[39] David Brash. Recent additions to the ARMv7-A architecture. In *Proc. of the 28th International IEEE Conference on Computer Design (ICCD)*, 2010. DOI: 10.1109/ICCD.2010.5647549 28, 123

[40] David Brash. The ARMv8-A architecture and its ongoing development, December 2014. http://community.arm.com/groups/processors/blog/2014/12/02/the-armv8-a-architecture-and-its-ongoing-development 143

[41] Jeffrey Buell, Daniel Hecht, Jin Heo, Kalyan Saladi, and H. Reza Taheri. Methodology for performance analysis of VMware vSphere under Tier-1 applications. *VMware Technical Journal*, 2(1), June 2013. 161

[42] Davidlohr Bueso. KVM: virtual x86 MMU setup. http://blog.stgolabs.net/2012/03/kvm-virtual-x86-mmu-setup.html, 2012. 72

[43] Edouard Bugnion, Vitaly Chipounov, and George Candea. Lightweight snapshots and system-level backtracking. In *Proc. of The 14th Workshop on Hot Topics in Operating Systems (HotOS-XIV)*, 2013. https://www.usenix.org/conference/hotos13/session/bugnion xvi

[44] Edouard Bugnion, Scott Devine, Kinshuk Govil, and Mendel Rosenblum. Disco: Running commodity operating systems on scalable multiprocessors. *ACM Transactions on Computer Systems*, 15(4):412–447, 1997. DOI: 10.1145/265924.265930 xiii, 29, 32, 34

[45] Edouard Bugnion, Scott Devine, Mendel Rosenblum, Jeremy Sugerman, and Edward Y. Wang. Bringing virtualization to the x86 architecture with the original VMware workstation. *ACM Transactions on Computer Systems*, 30(4):12, 2012. DOI: 10.1145/2382553.2382554 xiv, 7, 10, 34, 39, 40, 44, 51, 54, 66

[46] Edouard Bugnion, Scott W. Devine, and Mendel Rosenblum. System and method for virtualizing computer systems, September 1998. U.S. Patent 6,496,847. http://www.google.com/patents?vid=6496847 34, 40

[47] John Chapin, Mendel Rosenblum, Scott Devine, Tirthankar Lahiri, Dan Teodosiu, and Anoop Gupta. Hive: Fault containment for shared-memory multiprocessors. In *Proc. of the 15th ACM Symposium on Operating Systems Principles (SOSP)*, pages 12–25, 1995. DOI: 10.1145/224056.224059 30

[48] Peter M. Chen, Edward K. Lee, Garth A. Gibson, Randy H. Katz, and David A. Patterson. RAID: High-performance, reliable secondary storage. *ACM Computer Surveys*, 26(2):145–185, 1994. DOI: 10.1145/176979.176981 2

[49] Peter M. Chen and Brian D. Noble. When virtual is better than real. In *Proc. of the 8th Workshop on Hot Topics in Operating Systems (HotOS-VIII)*, pages 133–138, 2001. DOI: 10.1109/HOTOS.2001.990073 13

[50] David Chisnall. *The Definitive Guide to the Xen Hypervisor*. Prentice-Hall, 2007. 51, 69

[51] Christopher Clark, Keir Fraser, Steven Hand, Jacob Gorm Hansen, Eric Jul, Christian Limpach, Ian Pratt, and Andrew Warfield. Live migration of virtual machines. In *Proc. of the 2nd Symposium on Networked Systems Design and Implementation (NSDI)*, 2005. http://www.usenix.org/events/nsdi05/tech/clark.html xiv, 68, 80

[52] CloudLab. http://www.cloudlab.us 149

[53] Robert F. Cmelik and David Keppel. Shade: A fast instruction-set simulator for execution profiling. In *Proc. of the 1994 ACM SIGMETRICS International Conference on Measurement and Modeling of Computer Systems*, pages 128–137, 1994. DOI: 10.1145/183018.183032 38

[54] R. J. Creasy. The origin of the VM/370 time-sharing system. *IBM Journal of Research and Development*, 25(5):483–490, 1981. xvi

[55] Steve Dahl and Garry Meier. Disco demolition night. http://en.wikipedia.org/wiki/Disco_Demolition_Night, 1979. xvi

[56] Christoffer Dall, Jeremy Andrus, Alexander Van't Hof, Oren Laadan, and Jason Nieh. The design, implementation, and evaluation of cells: A virtual smartphone architecture. *ACM Transactions on Computer Systems*, 30(3):9, 2012. DOI: 10.1145/2324876.2324877 xvi

[57] Christoffer Dall and Andrew Jones. KVM/ARM unit tests. https://github.com/columbia/kvm-unit-tests 140

[58] Christoffer Dall, Shih-Wei Li, Jintack Lim, Jason Nieh, and Georgios Koloventzos. ARM virtualization: Performance and architectural implications. In *Proc. of the 43rd International Symposium on Computer Architecture (ISCA)*, June 2016. xvi, 147, 161

[59] Christoffer Dall and Jason Nieh. KVM for ARM. In *Proc. of the 12th Annual Linux Symposium*, Ottawa, Canada, July 2010. 26, 28, 47, 48, 51, 123

[60] Christoffer Dall and Jason Nieh. KVM/ARM: The design and implementation of the linux ARM hypervisor. In *Proc. of the 19th International Conference on Architectural Support for Programming Languages and Operating Systems (ASPLOS-XIX)*, pages 333–348, 2014. DOI: 10.1145/2541940.2541946 xvi, 123, 132, 146, 161

[61] Al Danial. cloc. https://github.com/AlDanial/cloc 142

[62] David Brash. Architecture program manager, ARM Ltd. Personal Communication, November 2012. 126

[63] Harvey M. Deitel. *An Introduction to Operating Systems*. Addison-Wesley, 1984. 82

[64] Peter J. Denning. The locality principle. *Communications of the ACM*, 48(7):19–24, 2005. DOI: 10.1145/1070838.1070856 76

[65] Scott W. Devine, Edouard Bugnion, and Mendel Rosenblum. Virtualization system including a virtual machine monitor for a computer with a segmented architecture, October 1998. U.S. Patent 6,397,242. http://www.google.com/patents?vid=6397242 34, 36

[66] Micah Dowty and Jeremy Sugerman. GPU virtualization on VMware's hosted I/O architecture. *Operating Systems Review*, 43(3):73–82, 2009. DOI: 10.1145/1618525.1618534 51

[67] Ulrich Drepper. The cost of virtualization. *ACM Queue*, 6(1):28–35, 2008. DOI: 10.1145/1348583.1348591 75

[68] George W. Dunlap, Samuel T. King, Sukru Cinar, Murtaza A. Basrai, and Peter M. Chen. ReVirt: Enabling intrusion analysis through virtual-machine logging and replay. In *Proc. of the 5th Symposium on Operating System Design and Implementation (OSDI)*, 2002. http://www.usenix.org/events/osdi02/tech/dunlap.html 13

[69] Roy T. Fielding and Gail E. Kaiser. The apache HTTP server project. *IEEE Internet Computing*, 1(4):88–90, 1997. DOI: 10.1109/4236.612229 113

[70] Brad Fitzpatrick. Distributed caching with memcached. *Linux Journal*, (124):5, Aug 2004. http://dl.acm.org/citation.cfm?id=1012889.1012894 113

[71] Bryan Ford and Russ Cox. Vx32: Lightweight user-level sandboxing on the x86. In *Proc. of the USENIX Annual Technical Conference (ATC)*, pages 293–306, 2008. http://www.usenix.org/events/usenix08/tech/full_papers/ford/ford.pdf 4, 29

[72] Jayneel Gandhi, Arkaprava Basu, Mark D. Hill, and Michael M. Swift. Efficient memory virtualization: Reducing dimensionality of nested page walks. In *Proc. of the 47th Annual IEEE/ACM International Symposium on Microarchitecture (MICRO)*, pages 178–189, 2014. DOI: 10.1109/MICRO.2014.37 77

[73] Tal Garfinkel, Ben Pfaff, Jim Chow, Mendel Rosenblum, and Dan Boneh. Terra: A virtual machine-based platform for trusted computing. In *Proc. of the 19th ACM Symposium on Operating Systems Principles (SOSP)*, pages 193–206, 2003. DOI: 10.1145/945445.945464 xiv, 13

[74] Pat Gelsinger. Personal Communication (Intel Corp. CTO), 1998. 25, 34

[75] General Dynamics. OKL4 Microvisor, February 2013. http://www.ok-labs.com/products/okl4-microvisor 146

[76] G. Benton Gibbs and Margo Pulles. Advanced POWER virtualization on IBM eServer p5 servers: Architecture and performance considerations. *IBM, International Technical Support Organization*, 2005. xv

[77] Robert P. Goldberg. *Architectural Principles for Virtual Computer Systems*. Ph.D. thesis, Harvard University, Cambridge, MA, 1972. http://www.dtic.mil/cgi-bin/GetTRDoc?AD=AD772809&Location=U2&doc=GetTRDoc.pdf 7

[78] Robert P. Goldberg. Survey of virtual machine research. *IEEE Computer Magazine*, 7(6):34–45, Jun 1974. xiii, xvi, 79

[79] Xiaoli Gong, Qi Du, Xu Li, Jin Zhang, and Ye Lu. Performance overhead of Xen on Linux 3.13 on ARM Cortex-A7. In *Proc. of the 9th International Conference on P2P, Parallel, Grid, Cloud and Internet Computing (3PGCIC)*, pages 453–456, 2014. DOI: 10.1109/3PGCIC.2014.92 161

[80] Abel Gordon, Nadav Amit, Nadav Har'El, Muli Ben-Yehuda, Alex Landau, Assaf Schuster, and Dan Tsafrir. ELI: Bare-metal performance for I/O virtualization. In *Proc. of the 17th International Conference on Architectural Support for Programming Languages and Operating Systems (ASPLOS-XVII)*, pages 411–422, 2012. DOI: 10.1145/2150976.2151020 xvii, 113, 114, 115

[81] Kinshuk Govil, Dan Teodosiu, Yongqiang Huang, and Mendel Rosenblum. Cellular disco: Resource management using virtual clusters on shared-memory multiprocessors. In *Proc. of the 17th ACM Symposium on Operating Systems Principles (SOSP)*, pages 154–169, 1999. DOI: 10.1145/319151.319162 29, 30

[82] Ajay Gulati, Irfan Ahmad, and Carl A. Waldspurger. PARDA: Proportional allocation of resources for distributed storage access. In *Proc. of the 7th USENIX Conference on File and Storage Technologie (FAST)*, pages 85–98, 2009. http://www.usenix.org/events/fast09/tech/full_papers/gulati/gulati.pdf 51

[83] Nadav Har'El, Abel Gordon, Alex Landau, Muli Ben-Yehuda, Avishay Traeger, and Razya Ladelsky. Efficient and scalable paravirtual I/O system. In *Proc. of the USENIX Annual Technical Conference (ATC)*, pages 231–242, 2013. https://www.usenix.org/conference/atc13/technical-sessions/presentation/har%E2%80%99el 121

[84] Kim M. Hazelwood. *Dynamic Binary Modification: Tools, Techniques, and Applications*. Synthesis Lectures on Computer Architecture. Morgan & Claypool Publishers, San Rafael, CA, 2011. 38

[85] John L. Hennessy, Norman P. Jouppi, Steven A. Przybylski, Christopher Rowen, Thomas R. Gross, Forest Baskett, and John Gill. MIPS: A microprocessor architecture. In *Proc. of the 15th Annual IEEE/ACM International Symposium on Microarchitecture (MICRO)*, pages 17–22, 1982. http://dl.acm.org/citation.cfm?id=800930 29

[86] Jin Heo and Reza Taheri. Virtualizing latency-sensitive applications: Where does the overhead come from? *VMware Technical Journal*, 2(2), December 2013. 161

[87] Hewlett-Packard. http://www8.hp.com/us/en/products/moonshot-systems/product-detail.html?oid=7398915, March 2015. 149

[88] Owen S. Hofmann, Sangman Kim, Alan M. Dunn, Michael Z. Lee, and Emmett Witchel. InkTag: Secure applications on an untrusted operating system. In *Proc. of the 18th International Conference on Architectural Support for Programming Languages and Operating Systems (ASPLOS-XVIII)*, pages 265–278, 2013. DOI: 10.1145/2451116.2451146 xvi

[89] The HSA Foundation. `http://www.hsafoundation.com/` 120

[90] The HSA Foundation. HSA-Drivers-Linux-AMD. `https://github.com/HSAFounda tion/HSA-Drivers-Linux-AMD`. Accessed: May 2016. 120

[91] J. Y. Hwang, S. B. Suh, S. K. Heo, C. J. Park, J. M. Ryu, S. Y. Park, and C. R. Kim. Xen on ARM: System virtualization using Xen hypervisor for ARM-based secure mobile phones. In *Proc. of the 5th Consumer Communications and Newtork Conference*, January 2008. 51, 123

[92] Ian Campbell. Personal Communication, April 2015. 159

[93] IBM Corporation. PowerLinux servers—64-bit DMA concepts. `http://pic.dhe.ib m.com/infocenter/lnxinfo/v3r0m0/topic/liabm/liabmconcepts.htm`. Accessed: August 2016. 102

[94] IBM Corporation. Taking advantage of 64-bit DMA capability on PowerLinux. `https://www.ibm.com/developerworks/community/wikis/home?lang=en#!/wi ki/W51a7ffcf4dfd_4b40_9d82_446ebc23c550/page/Taking%20Advantage%20of% 2064-bit%20DMA%20capability%20on%20PowerLinux`. Accessed: August 2016. 102

[95] IDC. Server virtualization hits inflection point as number of virtual machines to exceed physical systems (press release). `http://www.idc.com/about/viewpressrelease.jsp ?containerId=prUK21840309`, 2009. xiv, 13

[96] VMware Infrastructure. Resource management with vmware drs. *VMware Whitepaper*, 2006. 13

[97] InSignal Co. ArndaleBoard.org. `http://arndaleboard.org` 140

[98] Intel Corporation. DPDK: Data plane development kit. `http://dpdk.org`. Accessed: May 2016. 105

[99] Intel Corporation. Intel ethernet drivers and utilities. `https://sourceforge.net/pr ojects/e1000/`. Accessed: August 2016. 93, 94

[100] Intel Corporation. Intel 82540EM gigabit ethernet controller. `http://ark.intel.com/ products/1285/Intel-82540EM-Gigabit-Ethernet-Controller`, 2002. Accessed: August 2016. 93

[101] Intel Corporation. PCI/PCI-X family of gigabit ethernet controllers software developer's manual. Revision 4.0. `http://www.intel.com/content/dam/doc/manual/pci-pci-x-family-gbe-controllers-software-dev-manual.pdf`, March 2009. Accessed: August 2016. 93, 94

[102] Intel Corporation. Intel64 and IA-32 architectures software developer's manual. vol. 2 (2A and 2B). 2010. 38

[103] Intel Corporation. Intel64 and IA-32 architectures software developer's manual. vol. 3B: System programming guide, (part 2). 2010. 64

[104] Intel Corporation. Intel 64 and IA-32 architectures software developer's manual, 325462-044US, August 2012. 131, 154

[105] Intel Corporation. Intel ethernet controller XL710. `http://www.intel.com/content/dam/www/public/us/en/documents/datasheets/xl710-10-40-controller-datasheet.pdf`, 2016. Accessed: August 2016. 107

[106] Intel Corporation. Intel virtualization technology for directed I/O—architecture specification. Revision 2.4. `http://www.intel.com/content/dam/www/public/us/en/documents/product-specifications/vt-directed-io-spec.pdf`, June 2016. 79, 102, 103, 105, 116

[107] Intel Corporation. Intel64 and IA-32 architectures software developer's manual. vol. 3A: System programming guide, (part 1). Accessed: August 2016. 116, 118

[108] Rick A. Jones. Netperf: A network performance benchmark. Revision 2.0. `http://www.netperf.org/netperf/training/Netperf.html`, 1995. Accessed: August 2016. 97

[109] Asim Kadav, Matthew J. Renzelmann, and Michael M. Swift. Tolerating hardware device failures in software. In *Proc. of the 22nd ACM Symposium on Operating Systems Principles (SOSP)*, pages 59–72, 2009. DOI: 10.1145/1629575.1629582 105

[110] Poul-Henning Kamp and Robert N. M. Watson. Jails: Confining the omnipotent root. In *Proc. of the 2nd International SANE Conference*, vol. 43, page 116, 2000. 4

[111] Adrian King. *Inside Windows 95*. Microsoft Press, 1995. 36

[112] Avi Kivity. KVM unit tests. `https://git.kernel.org/cgit/virt/kvm/kvm-unit-tests.git` 140

[113] Avi Kivity. KVM: The linux virtual machine monitor. In *Proc. of the 2007 Ottawa Linux Symposium (OLS)*, pages 225–230, July 2007. `http://www.linuxsymposium.org/archives/OLS/Reprints-2007/kivity-Reprint.pdf` xiv, 7, 62, 66, 67, 132

[114] Teemu Koponen, Keith Amidon, Peter Balland, Martín Casado, Anupam Chanda, Bryan Fulton, Igor Ganichev, Jesse Gross, Paul Ingram, Ethan J. Jackson, Andrew Lambeth, Romain Lenglet, Shih-Hao Li, Amar Padmanabhan, Justin Pettit, Ben Pfaff, Rajiv Ramanathan, Scott Shenker, Alan Shieh, Jeremy Stribling, Pankaj Thakkar, Dan Wendlandt, Alexander Yip, and Ronghua Zhang. Network virtualization in multi-tenant datacenters. In *Proc. of the 11th Symposium on Networked Systems Design and Implementation (NSDI)*, pages 203–216, 2014. `https://www.usenix.org/conference/nsdi14/technical-sessions/presentation/koponen` 13

[115] Michael Kozuch and Mahadev Satyanarayanan. Internet suspend/resume. In *Proc. of the 4th IEEE Workshop on Mobile Computing Systems and Applications*, page 40, 2002. DOI: 10.1109/MCSA.2002.1017484 80

[116] Yossi Kuperman, Eyal Moscovici, Joel Nider, Razya Ladelsky, Abel Gordon, and Dan Tsafrir. Paravirtual remote I/O. In *Proc. of the 21st International Conference on Architectural Support for Programming Languages and Operating Systems (ASPLOS-XXI)*, pages 49–65, 2016. DOI: 10.1145/2872362.2872378 xvii, 121

[117] Jeffrey Kuskin, David Ofelt, Mark Heinrich, John Heinlein, Richard Simoni, Kourosh Gharachorloo, John Chapin, David Nakahira, Joel Baxter, Mark Horowitz, Anoop Gupta, Mendel Rosenblum, and John L. Hennessy. The Stanford FLASH multiprocessor. In *Proc. of the 21st International Symposium on Computer Architecture (ISCA)*, pages 302–313, 1994. DOI: 10.1109/ISCA.1994.288140 29

[118] The Linux kvm Project Homepage. `http://www.linux-kvm.org` 69

[119] George Kyriazis. Heterogeneous system architecture: A technical review. Technical report, AMD Inc., Aug 2012. Revision 1.0. `http://amd-dev.wpengine.netdna-cdn.com/wordpress/media/2012/10/hsa10.pdf`. Accessed: May 2016. 120

[120] Ilya Lesokhin, Haggai Eran, Shachar Raindel, Guy Shapiro, Sagi Grimberg, Liran Liss, Muli Ben-Yehuda, Nadav Amit, and Dan Tsafrir. Page fault support for network controllers. In *ACM International Conference on Architectural Support for Programming Languages and Operating Systems (ASPLOS)*, 2017. (to appear). xvii, 120

[121] Linaro Ubunty Trusty Images. `https://releases.linaro.org/14.07/ubuntu/trusty-images/server`, July 2014. 149

[122] Linux kernel 4.7 source code, drivers/iommu/intel-iommu.c (line 518). `http://lxr.free-electrons.com/source/drivers/iommu/intel-iommu.c?v=4.7#L518`. Accessed: August 2016. 105

[123] Linux ARM Kernel Mailing List. "tcp: Refine TSO autosizing" causes performance regression on Xen, April 2015. http://lists.infradead.org/pipermail/linux-arm-kernel/2015-April/336497.html 159

[124] Peter S. Magnusson, Magnus Christensson, Jesper Eskilson, Daniel Forsgren, Gustav Hållberg, Johan Högberg, Fredrik Larsson, Andreas Moestedt, and Bengt Werner. Simics: A full system simulation platform. *IEEE Computer*, 35(2):50–58, 2002. DOI: 10.1109/2.982916 5, 6

[125] Moshe Malka, Nadav Amit, Muli Ben-Yehuda, and Dan Tsafrir. rIOMMU: Efficient IOMMU for I/O devices that employ ring buffers. In *Proc. of the 20th International Conference on Architectural Support for Programming Languages and Operating Systems (ASPLOS-XX)*, pages 355–368, 2015. DOI: 10.1145/2694344.2694355 104

[126] Moshe Malka, Nadav Amit, and Dan Tsafrir. Efficient intra-operating system protection against harmful DMAs. In *Proc. of the 13th USENIX Conference on File and Storage Technologie (FAST)*, pages 29–44, 2015. https://www.usenix.org/conference/fast15/technical-sessions/presentation/malka xvii

[127] Alex Markuze, Adam Morrison, and Dan Tsafrir. True IOMMU protection from DMA attacks: When copy is faster than zero copy. In *Proc. of the 21st International Conference on Architectural Support for Programming Languages and Operating Systems (ASPLOS-XXI)*, pages 249–262, 2016. DOI: 10.1145/2872362.2872379 xvii

[128] Mellanox Technologies. Mellanox ConnectX-5 VPI adapter. http://www.mellanox.com/related-docs/user_manuals/ConnectX-5_VPI_Card.pdf, 2016. Accessed: August 2016. 107

[129] Dirk Merkel. Docker: Lightweight linux containers for consistent development and deployment. *Linux Journal*, (239):2, 2014. xvi, 4

[130] Microsoft Corporation. Virtual server 2005 R2 and hardware virtualization. https://blogs.msdn.microsoft.com/virtual_pc_guy/2006/05/01/virtual-server-2005-r2-and-%20hardware-virtualization/, 2006. xiv

[131] David S. Miller, Richard Henderson, and Jakub Jelinek. Dynamic DMA mapping guide. https://www.kernel.org/doc/Documentation/DMA-API-HOWTO.txt. Linux kernel documentation. Accessed: August 2016. 105

[132] Ingo Molnar. Hackbench. http://people.redhat.com/mingo/cfs-scheduler/tools/hackbench.c

[133] Antonios Motakis, Alexander Spyridakis, and Daniel Raho. Introduction on performance analysis and profiling methodologies for kvm on arm virtualization. *VLSI Circuits and Systems VI*, March 2013. 161

[134] Bhyrav Mutnury, Frank Paglia, James Mobley, Girish K. Singh, and Ron Bellomio. QuickPath interconnect (QPI) design and analysis in high speed servers. In *Topical Meeting on Electrical Performance of Electronic Packaging and Systems*, pages 265–268, 2010. http://dx.doi.org/10.1109/EPEPS.2010.5642789 82

[135] Michael Nelson, Beng-Hong Lim, and Greg Hutchins. Fast transparent migration for virtual machines. In *USENIX Annual Technical Conference*, pages 391–394, 2005. http://www.usenix.org/events/usenix05/tech/general/nelson.html xiv, 51, 68, 80

[136] Jason Nieh and Ozgur Can Leonard. Examining VMware. *Dr. Dobb's Journal*, 315:70–76, August 2000. xvi

[137] Jason Nieh and Chris Vaill. Experiences teaching operating systems using virtual platforms and linux. In *Proc. of the 36th SIGCSE Technical Symposium on Computer Science Education (SIGCSE)*, pages 520–524, 2005. DOI: 10.1145/1047344.1047508 xvi

[138] Steven Osman, Dinesh Subhraveti, Gong Su, and Jason Nieh. The design and implementation of zap: A system for migrating computing environments. In *Proc. of the 5th Symposium on Operating System Design and Implementation (OSDI)*, 2002. http://www.usenix.org/events/osdi02/tech/osman.html xvi

[139] PCI-SIG. Alternative routing-id interpretation (ARI). https://pcisig.com/sites/default/files/specification_documents/ECN-alt-rid-interpretation-070604.pdf, 2007. Accessed: August 2016. 108

[140] PCI-SIG. Address translation services. Revision 1.1. http://www.pcisig.com/specifications/iov/ats/, 2009. 120

[141] PCI-SIG. PCI express base specification. Revision 3.0. https://pcisig.com/specifications, 2010. Accessed: August 2016. 82, 86

[142] Omer Peleg, Adam Morrison, Benjamin Serebrin, and Dan Tsafrir. Utilizing the IOMMU scalably. In *Proc. of the USENIX Annual Technical Conference (ATC)*, pages 549–562, 2015. https://www.usenix.org/conference/atc15/technical-session/presentation/peleg xvii, 104

[143] Gerald J. Popek and Robert P. Goldberg. Formal requirements for virtualizable third generation architectures. *Communications of the ACM*, 17(7):412–421, 1974. DOI: 10.1145/361011.361073 xv, 6, 15, 18, 22, 25, 26, 27, 37, 56, 79, 128

[144] Shaya Potter and Jason Nieh. Apiary: Easy-to-use desktop application fault containment on commodity operating systems. In *Proc. of the USENIX Annual Technical Conference (ATC)*, 2010. https://www.usenix.org/conference/usenix-atc-10/apiary-easy-use-desktop-application-fault-containment-commodity-operating xvi

[145] Shaya Potter and Jason Nieh. Improving virtual appliance management through virtual layered file systems. In *Proc. of the 25th Large Installation System Administration Conference (LISA)*, 2011. https://www.usenix.org/conference/lisa11/improving-virtual-appliance-management-through-virtual-layered-file-systems xvi

[146] Ian Pratt, Keir Fraser, Steven Hand, Christian Limpach, Andrew Warfield, Dan Magenheimer, Jun Nakajima, and Asit Mallick. Xen 3.0 and the art of virtualization. In *Proc. of the 2005 Ottawa Linux Symposium (OLS)*, 2005. http://www.linuxsymposium.org/archives/OLS/Reprints-2005/pratt-Reprint.pdf 7, 51, 69

[147] George Prekas, Mia Primorac, Adam Belay, Christos Kozyrakis, and Edouard Bugnion. Energy proportionality and workload consolidation for latency-critical applications. In *Proc. of the 2015 ACM Symposium on Cloud Computing (SOCC)*, pages 342–355, 2015. DOI: 10.1145/2806777.2806848 xvi

[148] QEMU Networking Documentation. http://wiki.qemu.org/Documentation/Networking. Accessed: August 2016. 99

[149] Lars Rasmusson and Diarmuid Corcoran. Performance overhead of KVM on Linux 3.9 on ARM cortex-a15. *SIGBED Review*, 11(2):32–38, 2014. DOI: 10.1145/2668138.2668143 161

[150] Realtek Semiconductor Corp. RTL8139C(L)+—Advanced PCI/Mini-PCI/Cardbus 3.3V single-chip 10/100M fast ethernet controller. http://www.realtek.com.tw/products/productsView.aspx?Langid=1&PFid=6&Level=5&Conn=4&ProdID=17. Accessed: August 2016. 99

[151] John Scott Robin and Cynthia E. Irvine. Analysis of the intel pentium's ability to support a secure virtual machine monitor. In *Proc. of the 9th USENIX Security Symposium*, 2000. https://www.usenix.org/conference/9th-usenix-security-symposium/analysis-intel-pentiums-ability-support-secure-virtual 25, 27, 34, 44, 46, 54

[152] Phil Rogers. Heterogeneous system architecture (HSA): Overview and implementation. 2013. HC25. http://www.hotchips.org/wp-content/uploads/hc_archives/hc25/HC25.0T1-Hetero-epub/HC25.25.100-Intro-Rogers-HSA%20Intro%20HotChips2013_Final.pdf. Accessed: May 2016. 120

[153] Mendel Rosenblum, Edouard Bugnion, Scott Devine, and Stephen Alan Herrod. Using the SimOS machine simulator to study complex computer systems. *ACM Transactions on Modeling and Computer Simulation*, 7(1):78–103, 1997. DOI: 10.1145/244804.244807 5, 6, 36

[154] Rusty Russell. Virtio: Towards a de-facto standard for virtual I/O devices. *Operating Systems Review*, 42(5):95–103, 2008. DOI: 10.1145/1400097.1400108 96, 138, 148, 159

[155] Jerome H. Saltzer and M. Frans Kaashoek. *Principles of Computer Systems: An Introduction*. Morgan Kaufman, 2009. 14

[156] Jose Renato Santos, Yoshio Turner, G. John Janakiraman, and Ian Pratt. Bridging the gap between software and hardware techniques for I/O virtualization. In *Proc. of the 2008 USENIX Annual Technical Conference*, pages 29–42, 2008. 159

[157] Constantine P. Sapuntzakis, David Brumley, Ramesh Chandra, Nickolai Zeldovich, Jim Chow, Monica S. Lam, and Mendel Rosenblum. Virtual appliances for deploying and maintaining software. In *Proc. of the 17th Large Installation System Administration Conference (LISA)*, pages 181–194, 2003. http://www.usenix.org/publications/librar y/proceedings/lisa03/tech/sapuntzakis.html 13

[158] Arvind Seshadri, Mark Luk, Ning Qu, and Adrian Perrig. SecVisor: A tiny hypervisor to provide lifetime kernel code integrity for commodity OSes. In *Proc. of the 21st ACM Symposium on Operating Systems Principles (SOSP)*, pages 335–350, 2007. DOI: 10.1145/1294261.1294294 xvi

[159] Mark Silberstein, Bryan Ford, Idit Keidar, and Emmett Witchel. GPUfs: Integrating a file system with GPUs. In *Proc. of the 18th International Conference on Architectural Support for Programming Languages and Operating Systems (ASPLOS-XVIII)*, pages 485–498, 2013. DOI: 10.1145/2451116.2451169 121

[160] Standard Performance Evaluation Corporation. https://www.spec.org/jvm2008, March 2015.

[161] Udo Steinberg and Bernhard Kauer. NOVA: A microhypervisor-based secure virtualization architecture. In *Proc. of the EuroSys Conference*, pages 209–222, 2010. DOI: 10.1145/1755913.1755935 146

[162] Jeremy Sugerman, Ganesh Venkitachalam, and Beng-Hong Lim. Virtualizing I/O devices on VMware workstation's hosted virtual machine monitor. In *USENIX Annual Technical Conference*, pages 1–14, 2001. http://www.usenix.org/publications/library/pr oceedings/usenix01/sugerman.html xiv, 34, 40, 51

[163] Sun Microsystems. Beginner's guide to LDOMs: Understanding and deploying logical domains, 2007. xv

[164] Sun microsystems, Inc. UltraSPARC T2 supplement to the UltraSPARC architecture 2007 (draft D1.4.3). http://www.oracle.com/technetwork/systems/opensparc/t 2-14-ust2-uasuppl-draft-hp-ext-1537761.html, 2007. Accessed: August 2016. 102

[165] Andrew S. Tanenbaum and Herbert Bos. *Modern Operating Systems*, 4th ed. Prentice-Hall, 2014. 14

[166] Michael S. Tsirkin. Vhost_net: A kernel-level virtio server. `https://lwn.net/Articles/346267/`, August 2009. Accessed: August 2016. 97

[167] Michael S. Tsirkin. Vhost-net and virtio-net: Need for speed. In *KVM Forum*, 2010. `http://www.linux-kvm.org/images/8/82/Vhost_virtio_net_need_for_speed_2.odp` 100

[168] Michael S. Tsirkin, Cornelia Huck, Pawel Moll, and Rusty Russell. Virtual I/O device (virtio) version 1.0—committee specification 04. `http://docs.oasis-open.org/virtio/virtio/v1.0/cs04/virtio-v1.0-cs04.html`, March 2016. Accessed: August 2016. 96

[169] UEFI Forum. Advanced configuration and power interface specification version 6.1, January 2016. `http://www.uefi.org/sites/default/files/resources/ACPI_6_1.pdf` 82

[170] UEFI Forum. Unified extensible firmware interface specification version 2.6, January 2016. `http://www.uefi.org/sites/default/files/resources/UEFI%20Spec%202_6.pdf` 82

[171] Rich Uhlig, Gil Neiger, Dion Rodgers, Amy L. Santoni, Fernando C. M. Martins, Andrew V. Anderson, Steven M. Bennett, Alain Kägi, Felix H. Leung, and Larry Smith. Intel virtualization technology. *IEEE Computer*, 38(5):48–56, 2005. DOI: 10.1109/MC.2005.163 15, 27, 53, 57, 79

[172] Ronald C. Unrau, Orran Krieger, Benjamin Gamsa, and Michael Stumm. Hierarchical clustering: A structure for scalable multiprocessor operating system design. *Journal of Supercomputing*, 9(1/2):105–134, 1995. `http://www.springerlink.com/content/v25n5q66384n4384/fulltext.pdf` 30

[173] Prashant Varanasi and Gernot Heiser. Hardware-supported virtualization on ARM. In *Proc. of the Asia-Pacific Workshop on Systems (APSys)*, page 11, 2011. DOI: 10.1145/2103799.2103813 143, 146

[174] VDE—Virtual Distributed Ethernet. `http://vde.sourceforge.net/`. Accessed: August 2016. 99

[175] Ben Verghese, Scott Devine, Anoop Gupta, and Mendel Rosenblum. Operating system support for improving data locality on CC-NUMA compute servers. In *Proc. of the 7th International Conference on Architectural Support for Programming Languages and Operating Systems (ASPLOS-VII)*, pages 279–289, 1996. DOI: 10.1145/237090.237205 30

[176] VMware, Inc. VMware KB article 1021095: Transparent page sharing (TPS) in hardware MMU systems. `http://kb.vmware.com/kb/1021095`, 2014. 77

[177] Carl A. Waldspurger. Memory resource management in VMware ESX server. In *Proc. of the 5th Symposium on Operating System Design and Implementation (OSDI)*, 2002. `http://www.usenix.org/events/osdi02/tech/waldspurger.html` xiv, 7, 33, 46, 51, 75, 77

[178] Carl A. Waldspurger and Mendel Rosenblum. I/O virtualization. *Communications of the ACM*, 55(1):66–73, 2012. DOI: 10.1145/2063176.2063194 80, 99

[179] Jon Watson. Virtualbox: Bits and bytes masquerading as machines. *Linux Journal*, (166):1, 2008. 7

[180] Jon Watson. Virtualbox: Bits and bytes masquerading as machines. *Linux Journal*, (166), February 2008. `http://dl.acm.org/citation.cfm?id=1344209.1344210` 66

[181] Thomas F. Wenisch, Roland E. Wunderlich, Michael Ferdman, Anastassia Ailamaki, Babak Falsafi, and James C. Hoe. SimFlex: Statistical sampling of computer system simulation. *IEEE Micro*, 26(4):18–31, 2006. DOI: 10.1109/MM.2006.79 5

[182] Andrew Whitaker, Marianne Shaw, and Steven D. Gribble. Scale and performance in the denali isolation kernel. In *Proc. of the 5th Symposium on Operating System Design and Implementation (OSDI)*, 2002. `http://www.usenix.org/events/osdi02/tech/whitaker.html` xiv, 4, 12, 44

[183] Paul Willmann, Scott Rixner, and Alan L. Cox. Protection strategies for direct access to virtualized I/O devices. In *Proc. of the USENIX Annual Technical Conference (ATC)*, pages 15–28, 2008. `http://www.usenix.org/events/usenix08/tech/fullpapers/willman/willman.pdf` 105

[184] Emmett Witchel and Mendel Rosenblum. Embra: Fast and flexible machine simulation. In *Proc. of the ACM SIGMETRICS International Conference on Measurement and Modeling of Computer Systems*, pages 68–79, 1996. DOI: 10.1145/233013.233025 6, 38

[185] Rafal Wojtczuk and Joanna Rutkowska. Following the white rabbit: Software attacks against Intel VT-d technology. `http://invisiblethingslab.com/resources/2011/Software%20Attacks%20on%20Intel%20VT-d.pdf`, April 2011. Accessed: August 2016. 105

[186] David Woodhouse. Patchwork [1/7] iommu/vt-d: Introduce intel_iommu=pasid28, and pasid_enabled() macro. `https://patchwork.kernel.org/patch/7357051/`, October 2015. Accessed: August 2016. 105

[187] Xen Project. `http://wiki.xenproject.org/wiki/Tuning_Xen_for_Performance`, November 2015. 150

[188] Xen.org. Xen ARM. `http://xen.org/products/xen_arm.html` 132, 146

[189] Idan Yaniv and Dan Tsafrir. Hash, don't cache (the page table). In *Proc. of the ACM SIGMETRICS International Conference on Measurement and Modeling of Computer Systems*, pages 337–350, 2016. DOI: 10.1145/2896377.2901456 77

[190] Bennet Yee, David Sehr, Gregory Dardyk, J. Bradley Chen, Robert Muth, Tavis Ormandy, Shiki Okasaka, Neha Narula, and Nicholas Fullagar. Native client: A sandbox for portable, untrusted x86 native code. In *IEEE Symposium on Security and Privacy*, pages 79–93, 2009. DOI: 10.1109/SP.2009.25 4, 29

[191] Fengzhe Zhang, Jin Chen, Haibo Chen, and Binyu Zang. CloudVisor: Retrofitting protection of virtual machines in multi-tenant cloud with nested virtualization. In *Proc. of the 23rd ACM Symposium on Operating Systems Principles (SOSP)*, pages 203–216, 2011. DOI: 10.1145/2043556.2043576 22, 60

Authors' Biographies

EDOUARD BUGNION

Edouard Bugnion is a Professor in the School of Computer and Communication Sciences at EPFL in Lausanne, Switzerland. His areas of interest include operating systems, datacenter infrastructure (systems and networking), and computer architecture.

Before joining EPFL, Edouard spent 18 years in the U.S., where he studied at Stanford and co-founded two startups: VMware and Nuova Systems (acquired by Cisco). At VMware from 1998–2005, he played many roles including CTO. At Nuova/Cisco from 2005–2011, he helped build the core engineering team and became the VP/CTO of Cisco's Server, Access, and Virtualization Technology Group, a group that brought to market the Unified Computing System (UCS) platform for virtualized datacenters.

Together with his colleagues, Bugnion received the ACM Software System Award for VMware 1.0 in 2009. His paper on Disco received a Best Paper Award at SOSP '97 and was entered into the ACM SIGOPS Hall of Fame Award. At EPFL, he received the OSDI 2014 Best Paper Award for his work on the IX dataplane operating system. Bugnion has a Dipl.Eng. degree from ETH Zurich, an M.Sc. and a Ph.D. from Stanford University, all in computer science.

JASON NIEH

Jason Nieh is a Professor of Computer Science at Columbia University. He has made research contributions in software systems across a broad range of areas, including operating systems, virtualization, thin-client computing, cloud computing, mobile computing, multimedia, web technologies, and performance evaluation. Technologies he developed are now widely used in Android, Linux, and other major operating system platforms. Honors for his research work include the Sigma Xi Young Investigator Award, awarded once every two years in the physical sciences and engineering, a National Science Foundation CAREER Award, a Department of Energy Early Career Award, five IBM Faculty Awards and two IBM Shared University Research Awards, six Google Research Awards, and various best paper awards, including those from MobiCom, SIGCSE, SIGMETRICS, and SOSP. A dedicated teacher, he received the Distinguished Faculty Teaching Award from the Columbia Engineering School Alumni Association for his innovations in teaching operating systems and for introducing virtualization as a pedagogical tool. Nieh earned his B.S. from MIT and his M.S. and Ph.D. from Stanford University, all in electrical engineering.

DAN TSAFRIR

Dan Tsafrir is an Associate Professor at the Technion—Israel Institute of Technology. His research interests are focused on practical aspects of operating systems, hardware-software interactions, virtualization, security, and performance evaluation. Some of his research contributions were deployed in Linux and KVM. His work was featured in the Communications of the ACM research highlights section. He received the USENIX FAST best paper award, the IBM Pat Goldberg memorial best paper award (twice), the HiPEAC paper award (twice), the Klein research prize, the Henri Gutwirth award for outstanding research, and research/faculty awards from Google, IBM, Intel, Mellanox, and VMware. Tsafrir earned his B.Sc., M.Sc., and Ph.D. from the Hebrew University of Jerusalem, all in computing science (B.Sc. also in math). Before joining the Technion, he was a postdoctoral researcher at the IBM T.J. Watson Research Center, New York, in the the Advanced Operating Systems Group (K42) and the BlueGene System Software Group.

Index